T0212938

Lecture Notes in Computer Science　14632

Founding Editors

Gerhard Goos
Juris Hartmanis

Editorial Board Members

Elisa Bertino, *Purdue University, West Lafayette, IN, USA*
Wen Gao, *Peking University, Beijing, China*
Bernhard Steffen ⓘ, *TU Dortmund University, Dortmund, Germany*
Moti Yung ⓘ, *Columbia University, New York, NY, USA*

The series Lecture Notes in Computer Science (LNCS), including its subseries Lecture Notes in Artificial Intelligence (LNAI) and Lecture Notes in Bioinformatics (LNBI), has established itself as a medium for the publication of new developments in computer science and information technology research, teaching, and education.

LNCS enjoys close cooperation with the computer science R & D community, the series counts many renowned academics among its volume editors and paper authors, and collaborates with prestigious societies. Its mission is to serve this international community by providing an invaluable service, mainly focused on the publication of conference and workshop proceedings and postproceedings. LNCS commenced publication in 1973.

Thomas Stützle · Markus Wagner
Editors

Evolutionary Computation in Combinatorial Optimization

24th European Conference, EvoCOP 2024
Held as Part of EvoStar 2024
Aberystwyth, UK, April 3–5, 2024
Proceedings

Springer

Editors
Thomas Stützle
IRIDIA
Université libre de Bruxelles
Brussels, Belgium

Markus Wagner
Monash University
Clayton, VIC, Australia

ISSN 0302-9743   ISSN 1611-3349 (electronic)
Lecture Notes in Computer Science
ISBN 978-3-031-57711-6   ISBN 978-3-031-57712-3 (eBook)
https://doi.org/10.1007/978-3-031-57712-3

© The Editor(s) (if applicable) and The Author(s), under exclusive license
to Springer Nature Switzerland AG 2024

This work is subject to copyright. All rights are solely and exclusively licensed by the Publisher, whether the whole or part of the material is concerned, specifically the rights of translation, reprinting, reuse of illustrations, recitation, broadcasting, reproduction on microfilms or in any other physical way, and transmission or information storage and retrieval, electronic adaptation, computer software, or by similar or dissimilar methodology now known or hereafter developed.
The use of general descriptive names, registered names, trademarks, service marks, etc. in this publication does not imply, even in the absence of a specific statement, that such names are exempt from the relevant protective laws and regulations and therefore free for general use.
The publisher, the authors and the editors are safe to assume that the advice and information in this book are believed to be true and accurate at the date of publication. Neither the publisher nor the authors or the editors give a warranty, expressed or implied, with respect to the material contained herein or for any errors or omissions that may have been made. The publisher remains neutral with regard to jurisdictional claims in published maps and institutional affiliations.

This Springer imprint is published by the registered company Springer Nature Switzerland AG
The registered company address is: Gewerbestrasse 11, 6330 Cham, Switzerland

Paper in this product is recyclable.

Preface

Metaheuristics are high-level algorithmic strategies that provide valuable tools for quickly approximating high-quality solutions for combinatorial optimization problems and other types of optimization problems. Examples of metaheuristics are evolutionary algorithms, simulated annealing, tabu search, iterated local search and various other techniques and often these techniques are based on underlying efficient constructive and local improvement methods. Some of these metaheuristics are based on methods that are designed purely with the goal of avoiding the problems of local optima, which is the case of tabu search and iterated local search. Very often these techniques are inspired by techniques in the natural world, as is the case with evolutionary algorithms or simulated annealing. Although some of these techniques are based on the natural world, these techniques often differ quite a bit from these inspirations. Anyway, whatever the reason behind these metaheuristics, the active research behind these algorithms is seen by their wide range of applications, the increasing variety of complex optimization problems, new methods such as the inclusion of machine learning techniques and, at the same time, the increasing number of theoretical analyses of these techniques. The articles in this volume give a variety of these developments in combinatorial optimization, evolutionary algorithms and other metaheuristics, machine learning, and experimental and theoretical work.

This volume contains the proceedings of EvoCOP 2024, the 24th European Conference on Evolutionary Computation in Combinatorial Optimisation. The conference was held in the lovely city of Aberystwyth, UK from 3–5 April 2024. The EvoCOP conference series started in 2001, with the first workshop specifically devoted to evolutionary computation in combinatorial optimization, and it became an annual conference in 2004. EvoCOP 2024 was organized together with EuroGP (the 27th European Conference on Genetic Programming), EvoMUSART (the 13th International Conference on Artificial Intelligence in Music, Sound, Art and Design), and EvoApplications (the 27th International Conference on the Applications of Evolutionary Computation, formerly known as EvoWorkshops), in a joint event collectively known as EvoStar 2024. Previous EvoCOP proceedings were published by Springer in the *Lecture Notes in Computer Science* series (LNCS volumes 2037, 2279, 2611, 3004, 3448, 3906, 4446, 4972, 5482, 6022, 6622, 7245, 7832, 8600, 9026, 9595, 10197, 10782, 11452, 12102, 12692, 13222 and 13987). The table on the next page reports the statistics for each of the previous conferences.

This year, 12 out of 28 papers were accepted after a rigorous double-blind process with an average of just under four reviews per submission, resulting in a 43% acceptance rate. We would like to acknowledge the quality and timeliness of our high-quality and diverse Program Committee members' work. Each year the members give freely of their time and expertise, in order to maintain the high standards of EvoCOP and provide constructive feedback to help authors improve their papers. Decisions considered both the reviewers' reports and the evaluation of the program chairs. The 12 accepted papers cover a variety of topics, ranging from constructive algorithms, machine learning techniques ranging from neural network-based guidance to sparse surrogate models for optimization problems, the foundations of evolutionary computation algorithms and other search heuristics, to multi-objective optimization problems. Fundamental and methodological aspects deal with various questions related to the theory of metaheuristic algorithms, the question of where difficult instances are, and the study of metaheuristics' core components and their design. Applications cover problem domains such as routing and other permutation problems, graph coloring, MAX-SAT and others. We believe that the range of topics covered in this volume reflects the current state of research in the fields of metaheuristics and combinatorial optimization.

EvoCOP	LNCS vol.	Submitted	Accepted	Acceptance (%)
2024	14632	28	12	42.9
2023	13987	32	15	46.8
2022	13222	28	13	46.4
2021	12692	42	14	33.3
2020	12102	37	14	37.8
2019	11452	37	14	37.8
2018	10782	37	12	32.4
2017	10197	39	16	41
2016	9595	44	17	38.6
2015	9026	46	19	41.3
2014	8600	42	20	47.6
2013	7832	50	23	46
2012	7245	48	22	45.8
2011	6622	42	22	52.4
2010	6022	69	24	34.8
2009	5482	53	21	39.6
2008	4972	69	24	34.8
2007	4446	81	21	25.9
2006	3906	77	24	31.2
2005	3448	66	24	36.4
2004	3004	86	23	26.7
2003	2611	39	19	48.7
2002	2279	32	18	56.3
2001	2037	31	23	74.2

We would like to express our appreciation to the various persons and institutions making EvoCOP 2024 a successful event. Firstly, we thank the local organization team, led by Christine Zarges from Aberystwyth University in the UK. Our acknowledgments also go to SPECIES, the Society for the Promotion of Evolutionary Computation in Europe and its Surroundings. We extend our acknowledgments to Nuno Lourenço from the University of Coimbra, Portugal, for his dedicated work with the submission and the proceedings system, to João Correia from the University of Coimbra, Portugal, for the EvoStar publicity and social media service, to Zakaria Abdelmoiz Dahi from the University of Málaga, Spain, for managing the EvoStar website, and to Sérgio Rebelo, Jéssica Parente and João Correia from the University of Coimbra, Portugal, for their important graphic design work. We wish to thank our prominent keynote speakers, Sabine Hauert and Jon Timmis. Finally, we express our appreciation to Anna I. Esparcia-Alcázar from SPECIES, Europe, whose considerable efforts in managing and coordinating EvoStar helped towards building a unique, vibrant and friendly atmosphere.

Special thanks also to Christian Blum, Francisco Chicano, Carlos Cotta, Peter Cowling, Jens Gottlieb, Jin-Kao Hao, Jano van Hemert, Bin Hu, Arnaud Liefooghe, Manuel Lopéz-Ibáñez, Peter Merz, Martin Middendorf, Gabriela Ochoa, Luís Paquete, Leslie Pérez Cáceres, Günther R. Raidl, Sébastien Verel and Christine Zarges for their hard work and dedication at past editions of EvoCOP, making this one of the reference international events in evolutionary computation and metaheuristics.

April 2024   Thomas Stützle
                                                                                      Markus Wagner

Organization

Organizing Committee

Conference Chairs

Thomas Stützle   Université libre de Bruxelles, Belgium
Markus Wagner   Monash University, Australia

Local Organization

Christine Zarges   Aberystwyth University, UK

Publicity Chair

João Correia   University of Coimbra, Portugal

EvoStar Coordinator

Anna Esparcia-Alcázar   Universitat Politècnica de València, Spain

EvoCOP Steering Committee

Christian Blum	Artificial Intelligence Research Institute (IIIA-CSIC), Spain
Francisco Chicano	University of Málaga, Spain
Peter Cowling	Queen Mary University of London, UK
Jens Gottlieb	SAP AG, Germany
Jin-Kao Hao	University of Angers, France
Bin Hu	AIT Austrian Institute of Technology, Austria
Arnaud Liefooghe	University of Lille, France
Manuel Lopéz-Ibáñez	University of Manchester, UK
Martin Middendorf	University of Leipzig, Germany
Gabriela Ochoa	University of Stirling, UK
Luís Paquete	University of Coimbra, Portugal
Leslie Pérez Cáceres	Pontificia Universidad Católica de Valparaíso, Chile
Günther Raidl	Vienna University of Technology, Austria
Jano van Hemert	Optos, UK

Sébastien Verel	University of the Littoral Opal Coast, France
Christine Zarges	Aberystwyth University, UK

Society for the Promotion of Evolutionary Computation in Europe and Its Surroundings (SPECIES)

Penousal Machado (President)
Mario Giacobini (Secretary)
Francisco Chicano (Treasurer)

Program Committee

Thomas Bartz-Beielstein	Technische Hochschule Köln, Germany
Matthieu Basseur	Université du Littoral Côte d'Opale, France
Christian Blum	Artificial Intelligence Research Institute (IIIA-CSIC), Spain
Alexander Brownlee	University of Stirling, UK
Maxim Buzdalov	ITMO University, Russia
Arina Buzdalova	ITMO University, Russia
Christian Camacho-Villalón	Université libre de Bruxelles, Belgium
Josu Ceberio	University of the Basque Country, Spain
Marco Chiarandini	University of Southern Denmark, Denmark
Francisco Chicano	University of Málaga, Spain
Carlos Coello Coello	CINVESTAV-IPN, Mexico
Carlos Cotta	Universidad de Málaga, SPAIN
Nguyen Dang	University of St Andrews, UK
Bilel Derbel	University of Lille, France
Karl Doerner	University of Vienna, Austria
Carola Doerr	CNRS and Sorbonne University, France
Jonathan Fieldsend	University of Exeter, UK
Carlos M. Fonseca	University of Coimbra, Portugal
Alberto Franzin	Université Libre de Bruxelles, Belgium
Bernd Freisleben	Philipps-Universität Marburg, Germany
Carlos Garcia-Martinez	University of Córdoba, Spain
Adrien Goëffon	University of Angers, France
Jin-Kao Hao	University of Angers, France
Geir Hasle	SINTEF Digital, Norway
Mario Inostroza-Ponta	Universidad de Santiago de Chile, Chile
Ekhine Irurozki	Télécom Paris, France
Thomas Jansen	Aberystwyth University, UK
Andrzej Jaszkiewicz	Poznan University of Technology, Poland

Marie-Eleonore Kessaci	Université de Lille, France
Ahmed Kheiri	Lancaster University, UK
Frederic Lardeux	University of Angers, France
Johannes Lengler	ETH Zürich, Switzerland
Rhydian Lewis	Cardiff University, UK
Arnaud Liefooghe	University of Lille, France
Manuel Lopéz-Ibáñez	University of Manchester, UK
Jose A. Lozano	University of the Basque Country, Spain
Gabriel Luque	University of Málaga, Spain
Krzysztof Michalak	Wroclaw University of Economics and Business, Poland
Nysret Musliu	Vienna University of Technology, Austria
Gabriela Ochoa	University of Stirling, UK
Pietro Oliveto	University of Sheffield, UK
Beatrice Ombuki-Berman	Brock University, Canada
Luís Paquete	University of Coimbra, Portugal
Mario Pavone	University of Catania, Italy
Paola Pellegrini	Université Gustave Eiffel, France
Francisco B. Pereira	Polytechnic Institute of Coimbra, Portugal
Daniel Porumbel	Conservatoire National des Arts et Métiers, France
Abraham Punnen	Simon Fraser University, Canada
Günther Raidl	Vienna University of Technology, Austria
María Cristina Riff	Universidad Técnica Federico Santa María, Chile
Marcus Ritt	Universidade Federal do Rio Grande do Sul, Brazil
Eduardo Rodriguez-Tello	CINVESTAV – Tamaulipas, Mexico
Andrea Roli	Università di Bologna, Italy
Hana Rudová	Masaryk University, Czech Republic
Valentino Santucci	University of Perugia, Italy
Frederic Saubion	University of Angers, France
Kevin Sim	Edinburgh Napier University, UK
Giovanni Squillero	Politecnico di Torino, Italy
Dirk Sudholt	University of Passau, Germany
El-Ghazali Talbi	University of Lille, France
Sara Tari	Université du Littoral Côte d'Opale, France
Renato Tinós	University of São Paulo, Brazil
Nadarajen Veerapen	University of Lille, France
Sébastien Verel	Université du Littoral Côte d'Opale, France
Carsten Witt	Technical University of Denmark, Denmark
Christine Zarges	Aberystwyth University, UK
Fangfang Zhang	University of Wellington, New Zealand

Contents

A Neural Network Based Guidance for a BRKGA: An Application to the Longest Common Square Subsequence Problem

Jaume Reixach[1]([✉])[ID], Christian Blum[1][ID], Marko Djukanović[2][ID],
and Günther R. Raidl[3][ID]

[1] Artificial Intelligence Research Institute (IIIA-CSIC), Campus of the UAB,
08193 Bellaterra, Spain
{jaume.reixach,christian.blum}@iiia.csic.es
[2] Faculty of Natural Sciences and Mathematics, University of Banja Luka,
Mladena Stojanovića 2, 78000 Banja Luka, Bosnia and Herzegovina
marko.djukanovic@pmf.unibl.org
[3] Institute of Logic and Computation, TU Wien, Vienna, Austria
raidl@ac.tuwien.ac.at

Abstract. In this work we apply machine learning to better guide a biased random key genetic algorithm (BRKGA) for the longest common square subsequence (LCSqS) problem. The problem is a variant of the well-known longest common subsequence (LCS) problem in which valid solutions are square strings. A string is square if it can be expressed as the concatenation of a string with itself. The original BRKGA is based on a reduction of the LCSqS problem to the LCS problem by cutting each input string into two parts. Our work consists in enhancing the search process of BRKGA for good cut points by using a machine learning approach, which is trained to produce promising cut points for the input strings of a problem instance. In this study, we show the benefits of this approach by comparing the enhanced BRKGA with the original BRKGA, using two benchmark sets from the literature. We show that the results of the enhanced BRKGA significantly improve over the original results, especially when tackling instances with non-uniformly generated input strings.

Keywords: Genetic algorithms · Neural networks · Longest Common Subsequences · Beam search

Jaume Reixach and Christian Blum are supported by grants TED2021-129319B-I00 and PID2022-136787NB-I00 funded by MCIN/AEI/10.13039/501100011033. Günter R. Raidl is supported by the Vienna Graduate School on Computational Optimization (VGSCO), Austrian Science Foundation, project no. W1260-N35. Marko Djukanović is supported by the project entitled "Development of artificial intelligence models and algorithms for solving difficult combinatorial optimization problems" funded by the Ministry of Scientific and Technological Development and the Higher Education of the Republic of Srpska.

© The Author(s), under exclusive license to Springer Nature Switzerland AG 2024
T. Stützle and M. Wagner (Eds.): EvoCOP 2024, LNCS 14632, pp. 1–15, 2024.
https://doi.org/10.1007/978-3-031-57712-3_1

1   Introduction

Recently, the use of machine learning (ML) for guiding metaheuristic algorithms has become popular in order to enhance performance [2]. In this work, we show how this idea can be beneficially applied within genetic algorithms (GAs). More specifically, information obtained through a ML approach is used to guide a biased random key genetic algorithm (BRKGA). In the context of combinatorial optimization, BRKGA work on a population of individuals indirectly representing solutions to the problem at hand by means of vectors of real values in $(0, 1)$. An important aspect of a BRKGA is the so-called decoder which maps every individual to a solution to the problem at hand. The methodology proposed in this work consists of biasing individuals towards vectors *learned* by a feed-forward neural network [1] before applying the decoder.

Our approach is evaluated using the Longest Common Square Subsequence (LCSqS) problem, a variation of the well-known Longest Common Subsequence (LCS) problem. Both problems are formally introduced in the following.

1.1   The LCSqS Problem

A string is considered a finite sequence of characters drawn from a finite set Σ, referred to as the alphabet. Given a string s, a subsequence of s is a string that can be derived from s by selectively removing zero or more characters while preserving the original order of the remaining ones. The longest common subsequence (LCS) problem entails, given a set of input strings $S = \{s_1, s_2, \ldots, s_m\}$ ($m \geq 2$), the task of identifying the longest possible string that is as a subsequence of all the strings in S. This problem, known to be NP-hard when the number of input strings is not fixed [13], finds vital applications across diverse domains including bioinformatics, file plagiarism detection, and time series analysis [11,14,15,18].

In our present study, we concentrate on a specific variation of this problem known as the longest common square subsequence (LCSqS) problem as introduced by Inoue et al. [9]. A string s is classified as a square string if it can be expressed by concatenating a string s' with itself. The aim of the LCSqS problem is to find a longest common subsequence within the set S of input strings that is also a square string. Similarly to the LCS problem, the LCSqS problem possesses applications in bioinformatics, in particular facilitating the identification of internal structural similarities within molecular data [8].

Reduction to the LCS Problem. Before we explain the reduction, let us introduce additional notation. For a string s, we denote its length by $|s|$. For two integers $i, j \leq |s|$, $s[i, j]$ refers to a (continuous) part of string s that starts from the character at position i and ends with the character at position j; when $i = j$, the single-character string $s[i]$ is given, or when $i > j$, the empty string ε. Note that the starting character of each string holds position one.

A notable characteristic of the LCSqS problem, effectively leveraged by existing heuristic algorithms, is its reducibility to the LCS problem. Given a set of input strings $S = \{s_1, s_2, \ldots, s_m\}$, let $\mathcal{P} = \{(p_1, p_2, \ldots, p_m) \in \prod_{i=1}^{m}\{1, \ldots, |s_i| - 1\}\}$ denote the set of all possibilities for partitioning each string of S into two parts. The LCSqS problem with input strings S can then be solved as follows. First, for every $p \in \mathcal{P}$ a solution s_p to the LCS problem with input strings $S_p = \{s_1[1, p_1], s_1[p_1 + 1, |s_1|], s_2[1, p_2], s_2[p_2 + 1, |s_2|], \ldots, s_m[1, p_m], s_m[p_m + 1, |s_m|]\}$ is computed. It is rather easy to see that the concatenation of $s_p^* = \arg\max_{p \in \mathcal{P}} s_p$ with itself, that is $s_p^* \cdot s_p^*$, gives an optimal solution to the original LCSqS problem [16]. Thus, the LCSqS problem with input strings S can be solved by finding the cut point vector $p \in \mathcal{P}$ that leads to the longest LCS solution for input strings S_p and concatenating this LCS solution with itself.

1.2   Literature Review

Numerous algorithms have been devised for addressing the LCS problem, owing to its many important practical applications. Exact solutions can be obtained through dynamic programming approaches, but their computational complexity is in $O(n^m)$, where m refers to the number of input strings, and n is the length of the longest input string. As this number of strings m grows, the practicality of these dynamic programming methods diminishes, leading to the adoption of heuristic and metaheuristic approaches. Among these, one of the most successful is beam search (BS), which was initially introduced in the context of the LCS problem by Blum et al. in [3].

When it comes to the LCSqS problem, Inoue et al. introduced exact dynamic programming algorithms in [9], which necessitate $O(n^6)$ time for the scenario involving two input strings. Conversely, Djukanović et al. [8] proposed two heuristic algorithms, with the most effective one being a hybrid approach combining reduced variable neighborhood search (RVNS) with BS. The so far leading heuristic algorithm is a BRKGA searching through the space of cut points and using BS for solving the corresponding LCS problems [16]. The used BS approach is the one outlined by Djukanovíc et al. [7], which offers two distinct designs for its guiding heuristic function. In our work, we enhance this BRKGA through the integration of a ML technique, and we refer to this enhanced version as BRKGA-LEARN. A brief overview of BRKGA is provided in the following sections, while a more comprehensive explanation of this algorithm can be found in the referenced articles.

1.3   Our Contribution

As already mentioned, the original BRKGA for the LCSqS problem from [16] searches the space of possible cut points heuristically, because a complete exploration of such a vast space is intractable in general. Each possible individual is mapped to a valid solution (consisting of a cut point for each input string) and then evaluated by applying the BS approach from [7] to the resulting LCS problem. Due to the high complexity of solving the LCS problem, this step of solving each LCS problem instance is done in a heuristic manner by BS. The

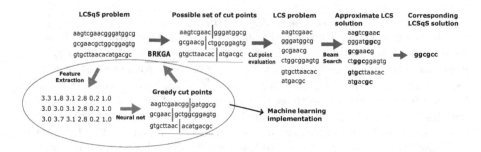

Fig. 1. General structure of the proposed BRKGA-LEARN approach

mapping process is performed within a decoder that also makes use of greedy information consisting of cut points deemed promising for every input string.

We introduce a novel approach to enhance the BRKGA for the LCSqS problem from [16] by replacing its original greedy mechanism with *learned* information provided by a ML model. Specifically, we advocate the utilization of a feed-forward *neural network* (NN). To determine a cut point for a given string s from the set S of all input strings, this NN takes as input features from s in combination with overarching features drawn from the entire set S. The overall structure of our proposed approach is illustrated in Fig. 1. First, given a set of input strings for the LCSqS problem, individual features for each input string and global features of S are extracted. Afterward, these are fed into the (previously trained) neural net, obtaining a cut point for each input string. Finally, these cut points are used by the decoder of the BRKGA, which uses them in order to replace greedy information.

The rest of this work is organized as follows. In Sect. 2, the original BRKGA is explained with a focus on its cut point search. Next, Sect. 3 develops the proposed learning mechanism, explaining the selected features and the training approach used for the NN. Section 4 presents a comprehensive experimental evaluation including a comparison of the proposed BRKGA-LEARN with the original BRKGA. Finally, in Sect. 5 some conclusions are derived and an outlook on future work is given.

2   The Original BRKGA

The top part of Fig. 1 illustrates the main idea of the BRKGA for the LCSqS problem. In the following, we will summarize the working mechanism of this algorithm. For an explanation of the BS we refer to the original publication [7].

In the context of any combinatorial optimization problem, a BRKGA works on a population in which each individual **v**, represented by a vector of real values from $(0, 1)$, can be mapped to a valid solution to the problem at hand. In the case of the application to the LCSqS problem, an individual **v** is mapped to a cut point vector for the set of input strings S, that is, an element of the set of all possible cut point vectors $\mathcal{P}$ as defined in Sect. 1.1. The goal is to find an

---

Algorithm 1. The decoder of the BRKGA for the LCSqS problem

Input: Input strings $S = \{s_1, \ldots, s_m\}$, an individual $\mathbf{v}$, beam width β, and BS guidance function h

1: $\mathbf{v}' \leftarrow$ greedy_transformation($\mathbf{v}$)
2: $\mathbf{p}^{\mathbf{v}} \leftarrow$ map_to_cut_points($\mathbf{v}'$)
3: $S_{\mathbf{p}^{\mathbf{v}}} \leftarrow$ LCS problem instance induced by $\mathbf{p}^{\mathbf{v}}$
4: $t^{\mathbf{v}} \leftarrow$ beam_search_for_LCS_problem($S_{\mathbf{p}^{\mathbf{v}}}, \beta, h$)
5: **return** $t^{\mathbf{v}} \cdot t^{\mathbf{v}}$

---

individual $\mathbf{v}$ that maps to a cut point vector $\mathbf{p}^{\mathbf{v}} \in P$ that maximizes the length of the solution to the LCS problem with input strings $S_{\mathbf{p}^{\mathbf{v}}}$. The decoder that takes care of the mapping process is a critical part of the algorithm and will be explained below.

A BRKGA works as follows. First, a population of p_{size} individuals is initialized at random, by setting every individual to a random vector of values from $(0, 1)$. Afterward, the main loop is entered, in which the fitness of each individual is determined by applying the decoder and further evaluating the obtained solution. The population is then split into the following two parts:

1. The *elite* population $P_e \subset P$ that consists of the best p_e individuals of P.
2. The *non-elite* population, consisting of the remaining $p_{\text{size}} - p_e$ individuals of the current population P.

Here, $p_e < p_{\text{size}} - p_e$ is a algorithm parameter, called the number of elites. Another algorithm parameter, $p_m < p_{\text{size}} - p_e$, called the number of mutants, is then used to generate the next population of individuals in the following way. The elite population is passed to the next generation along with p_m mutant individuals, which are constructed randomly as in the case of the initial population. The remaining $p_{\text{size}} - p_e - p_m$ individuals are introduced through the process of mating. This consists of selecting two parents at random from the current population, one elite and one non-elite, and constructing a new individual by setting its i-th vector position to one of the parents' i-th vector positions, choosing for each position between the two parents depending on a parameter $\rho_e \in (0.5, 1]$, called the elite inheritance probability, which determines the probability of choosing the i-th vector position of the elite parent.

2.1 The Decoder

The decoder, which is shown in Algorithm 1, first applies a greedy transformation to individual $\mathbf{v}$ (line 1). Hereby, some greedy information is given by a vector $\mathbf{u}$ of the same dimension as $\mathbf{v}$ and the following expression is applied:

$$v_i' := v_i + \gamma \cdot (u_i - v_i) \tag{1}$$

Algorithm parameter $\gamma \in [0, 1]$ is the so-called *greedy rate* and controls the extent to which v_i is moved towards u_i, for all $i = 1, \ldots, m$. In case $\gamma = 1$, v_i

is simply replaced by u_i. In the other extreme, if $\gamma = 0$ the greedy information is not used at all. In [16], three different designs were proposed for u (see the subsequent section). Next, vector $\mathbf{v}' = (v'_1, v'_2, \ldots, v'_m) \in (0,1)^m$ is mapped to the cut point vector $\mathbf{p}^v$, where $p_i^v = \lfloor v'_i \cdot |s_i| \rceil$ for all $i = 1, \ldots, m$; see line 2 of Algorithm 1. Notation $\lfloor r \rceil$ refers to rounding value r to the closest integer. If p_i^v is 0 or $|s_i|$, the cut is set to 1 and $|s_i| - 1$ respectively, in order to only consider feasible cuts. Finally, BS is applied to the $2m$ input strings obtained after cutting the strings of S at the positions given by $\mathbf{p}^v$. The resulting LCS solution is concatenated with itself, producing a solution to the original LCSqS problem. After this is done, a measure of fitness is given to the individual, depending on its associated solution quality. The fitness measure used comprises two distinct values. The first value represents the length of the solution, while the second value is designed as a tie breaker to differentiate between individuals generating solutions of equal length. For a comprehensive description of this secondary fitness measure, which is elaborated upon with three different designs, we refer to the original article [16].

2.2 Different Designs for the Greedy Vector u

1. The first option consists in simply using $\mathbf{u} = (0.5, \ldots, 0.5)$. This is motivated by the fact that, in general, the middle point of every input string is potentially a good place for cutting the string into two pieces, as these cuts maximize the resulting strings minimum length obtained after cutting.
2. The second approach determines u_i for all $i = 1, \ldots, m$ by

$$u_i = \arg\min_{r \in [0,1]} \sum_{a \in \Sigma} \left| \left| s_i[1, \lfloor r \cdot |s_i| \rceil] \right|_a - \left| s_i[\lfloor r \cdot |s_i| \rceil + 1, |s_i|] \right|_a \right| \tag{2}$$

Here, $\left| s_i[1, \lfloor r \cdot |s_i| \rceil] \right|_a$ and $\left| s_i[\lfloor r \cdot |s_i| \rceil + 1, |s_i|] \right|_a$ denote the numbers of occurrences of character a in the two strings obtained after cutting s_i after position $\lfloor r \cdot |s_i| \rceil$. This approach looks for the cut that maximizes the overall equilibrium of the quantity of each character at both sides of the cut. The value that minimizes the previous expression may not be unique. In this case a random value among the candidate values is chosen. With this design, the greedy value exploits some information about the distribution of characters within the strings to decide which cut appears most promising.
3. The last considered design determines u_i for all $i = 1, \ldots, m$ by

$$u_i = \arg\max_{r \in [0,1]} \left| \mathrm{LCS}\left(s_i[1, \lfloor r \cdot |s_i| \rceil], \ s_i[\lfloor r \cdot |s_i| \rceil + 1, |s_i|] \right) \right| \tag{3}$$

Hereby, the LCS of the pair of substrings is determined with the dynamic programming approach from [17]. The motivation for this design is that the ultimate goal for deciding the cuts is maximizing the LCS length between the resulting $2\,m$ strings. As it is intractable to maximize this value for all strings together, we choose each cut to be the one that maximizes the LCS length between the resulting two parts of each string. Just as with the second design, the value that maximizes the latter expression may not be unique. In this case, the tie is broken randomly.

3   Machine Learning Based Guidance for BRKGA

Our approach presented in this paper, henceforth called BRKGA-LEARN, differs from the original BRKGA in the design of the greedy information vector u. More specifically, u_i is now determined by a feed-forward NN that receives features of the input string s_i together with some global features.

3.1   Features

Six features are used as input to the NN for each input string. The first two are specific to each string, while the last four are global ones. With this, we intend to capture information both about individual strings and the whole problem instance. Given a problem instance consisting of input strings $S = \{s_1, s_2, \ldots, s_m\}$, we denote by $\mathrm{gv2}_i$ and $\mathrm{gv3}_i$ the values for the second and third greedy value designs as outlined in the previous sub-section for the i-th string respectively.

The following features are extracted for every string s_i, $i = 1, 2, \ldots, m$:

$$X = \left(\mathrm{gv2}_i, \ \mathrm{gv3}_i, \ \overline{\mathrm{gv2}}, \ \overline{\mathrm{gv3}}, \ \sigma(\mathrm{gv2}), \ \sigma(\mathrm{gv3}) \right)$$

Hereby, $\overline{\mathrm{gv2}}$ and $\overline{\mathrm{gv3}}$ denote the averages of the second and third greedy values across all strings in S and $\sigma(\mathrm{gv2})$ and $\sigma(\mathrm{gv3})$ are the corresponding sample standard deviations, respectively:

$$\overline{\mathrm{gv2}} = \frac{\sum_{i=1}^{m} \mathrm{gv2}_i}{m}, \ \sigma(\mathrm{gv2}) = \sqrt{\frac{\sum_{i=1}^{m} \left(\mathrm{gv2}_i - \overline{\mathrm{gv2}} \right)^2}{m-1}} \tag{4}$$

$$\overline{\mathrm{gv3}} = \frac{\sum_{i=1}^{m} \mathrm{gv3}_i}{m}, \ \sigma(\mathrm{gv3}) = \sqrt{\frac{\sum_{i=1}^{m} \left(\mathrm{gv3}_i - \overline{\mathrm{gv3}} \right)^2}{m-1}} \tag{5}$$

Feature values are standardized before being fed into the NN in order to have a mean of zero and a standard deviation of one. The NN comprises a single node in the output layer with a sigmoid activation function as the NN is applied to a single string at a time, representing a promising cut point in the form of a value in $(0, 1)$. Moreover, tuning indicated that one dense hidden layer consisting of just five nodes with a ReLU activation function is sufficient, as more complex networks did not yield significantly better results. Figure 2 provides a graphical representation of the used NN architecture.

The expressions for the ReLU and the sigmoid function are given below.

$$\mathrm{ReLU}(x) = \max\{0, x\}, \ x \in \mathbb{R} \tag{6}$$

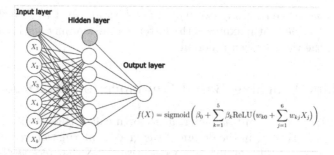

Fig. 2. A graphical representation of the employed feed-forward NN. $f(X)$ is the output of the neural network for the feature value vector $X = (X_1, X_2, \ldots, X_6)$. $\{w_{k,j}\}$ and $\{\beta_k\}$ are the 41 parameters of the neural net. Particularly, $\{w_{k,j}\}$ transform the features into the hidden-layer and $\{\beta_k\}$ transform the hidden-layer into the output. Morever, the lines from the top node in each layer represent the biases, which consist of parameters $\{w_{k0}\}$ and β_0.

$$\text{sigmoid}(x) = \frac{1}{1 + e^{-x}}, \quad x \in \mathbb{R} \tag{7}$$

3.2 Neural Net Training

To train the NN we have considered the following two options:

1. The first option consists in generating toy LCSqS problem instances, finding the optimal cuts for their input strings by complete enumeration, and then using these to train the NN. With this method each string and optimal cut pair is used as a training example for the NN, which is then trained in a classical supervised fashion using the Adam [10] stochastic gradient descent-based optimizer. A critical question clearly is here whether or not such a NN will generalize to out-of-distribution inputs for larger problem instances.
2. Training the NN on full-size instances using a Genetic Algorithm (GA). Hereby, each individual represents a complete set of weights for the NN, and the population is evolved until overfitting is detected. This has been shown to be a valid alternative for training neural networks [5,6].

Overall, the first approach produced unsatisfactory results. This may be attributed to the limitation that finding optimal cuts through complete enumeration is only feasible for problem instances of modest size. In our study, we employed instances, each comprising five strings of length 50. It is likely that the model's inability to generalize effectively to larger instances was due to the small size of the training data. As a result, we decided to pursue the second approach, which does not depend on knowledge of the optimal cuts for the training instances.

To delve into further detail, the genetic algorithm (GA) utilized for the NN training is a random key genetic algorithm (RKGA), which operates very similarly to the general process employed in a BRKGA. The primary distinction lies in the mating procedure. In the case of a RKGA, parental candidates are chosen randomly, whereas in a BRKGA, one parent is selected from the elites, and the other from the non-elite group. Additionally, in an RKGA, the offspring is derived by randomly selecting positions from the parents' vectors. The RKGA was applied using a population of 20 individuals, one elite individual, and seven mutants. Remember, in this context, that each individual is a vector of 41 real values that represent the NN weights.

Given a set of training LCSqS instances, the evaluation of an individual works as follows. First, a value u_i is obtained by supplying the NN (equipped with the weights of the individual) for every string s_i of every problem instance in the training set. Afterward, the decoder is applied in order to obtain the corresponding cut points. Lastly, BS is used to evaluate the obtained cut points. The average of the obtained solution lengths for all training instances is then used as a measure of fitness and associated to the corresponding individual. One may object that it would be better to run the whole BRKGA instead of just BS in order to evaluate an individual. However, this would lead to training times too large for practical purposes.

Once a new best individual is found during the search process of the training GA, a validation value is calculated in order to check for overfitting. This is done on the basis of a set of validation instances. The validation value is calculated by applying BRKGA for 10 min on the validation instances guided by the u vector obtained by the NN equipped with the corresponding individuals' weights. The average of the resulting LCSqS lengths is used as a validation value. Note that, in the case of validation it is computationally feasible to run BRKGA instead of BS, as new best individuals are found only sporadically.

Finally, when designing the training procedure one has to decide on an allowed domain for values for the NN weights. This is crucial since, during the generation of random individuals, they must be assigned random values within a specific interval. In our case we allowed weights to take real values from $(-1, 1)$. When allowing for larger values we observed that the NN fitted with random weights often produced output values close to either 0 or 1, which is undesirable as cut points too close to the start or end of strings usually lead to poor results.

In the BS for the NN training, we always applied a beam width of $\beta = 250$ and the UB$_1$ guidance function from [3]. For BRKGA, we used the parameter values determined for the experimental evaluation in the original BRKGA article, which were obtained by tuning depending on the benchmark set and string length.

4   Experimental Evaluation

In this section we experimentally evaluate BRKGA-LEARN and compare its performance to the one of the original BRKGA, the so far leading approach for solving the LCSqS problem. To do so, we train and run BRKGA-LEARN on two benchmark sets that were already used for the evaluation of the original BRKGA in [16]. The first benchmark set, named RANDOM, consists of instances with strings generated uniformly at random. In contrast, the second set named NON-RANDOM consists of instances with non-uniform strings generated by implanting appropriate patterns. For evaluating the original BRKGA a third benchmark set consisting of real-world instances was used, which we do not consider here due to its strong structural similarity to benchmark set RANDOM. For a more detailed explanation of the benchmark sets we refer to [16].

In order to keep the comparison fair, BRKGA-LEARN was given the same computation time for the application to each problem instance as BRKGA. In particular, we used a computation time limit of 600 CPU seconds per instance. This time limit was also the one applied for each algorithm execution during parameter tuning and for the calculation of the validation value during training. Benchmark sets RANDOM and NON-RANDOM consist of 150 and 100 instances, respectively, for each $n \in \{100, 500, 1000\}$, where n denotes the length of the strings in the instances. Moreover, NON-RANDOM instances can be split into two sets depending on a parameter $type$, which indicates the way in which the patterns were implanted.

BRKGA-LEARN was trained depending on n for instance set RANDOM and depending on n and $type$ for instance set NON-RANDOM. Therefore, three separate training procedures were conducted for benchmark set RANDOM, and six ones for benchmark set NON-RANDOM. On the other hand, we performed the parameter tuning depending on n for both instance sets, following the same procedure as for the original BRKGA. Moreover, we also used different equivalently generated instances for training, parameter tuning, and evaluation. Each RANDOM training used fifteen instances for calculating training values and fifteen more for calculating validation values. One for every combination of m (number of strings) and $|\Sigma|$ (alphabet size). Similarly, NON-RANDOM trainings used two sets of ten instances, with two instances for every possible value of m.

BRKGA-LEARN was implemented in C++ and training was executed in parallel using the OpenMP API [4], with the goal of speeding up the training process. Each training and evaluation run was executed on a cluster of machines with 10-core Intel Xeon processors at 2.2Ghz with 8Gb of RAM. The parallelism in the training runs was implemented in the calculation of the validation and training values. Each training uses 10 cores by distributing the training and validation instances within these. On the other hand, no parallelism was used for parameter tuning and the final experimental evaluation, just like in the case of the original BRKGA. Early stopping was used for the training runs, meaning that they were run until the validation value decreased, which indicates a possible overfitting. Training runs lasted about one hour on average, with runs trained using larger instances requiring up to four hours.

Tables 1 and 2 show the results obtained from parameter tuning. Firstly, p_e, p_m, ρ_e, p_{size}, $of2$ are the proportion of elites, the proportion of mutants, the elite inheritance probability, the population size and the secondary objective function design, respectively. Secondly, γ is the greedy rate, which controls the amount of greedy information used, and finally, β and h are the parameters of the beam search, namely the beam width and the guiding function design.

Table 1. Parameter tuning results for benchmark set RANDOM.

	p_e	p_m	ρ_e	p_{size}	$of2$	γ	β	h
$n = 100$	0.15	0.13	0.35	706	1	0.51	113	UB$_1$
$n = 500$	0.21	0.03	0.45	329	3	0.92	53	UB$_1$
$n = 1000$	0.31	0.28	0.68	737	2	0.98	587	UB$_1$

Table 2. Parameter tuning results for benchmark set NON-RANDOM.

	p_e	p_m	ρ_e	p_{size}	$of2$	γ	β	h
$n = 100$	0.18	0.04	0.47	906	2	0.79	388	UB$_1$
$n = 500$	0.16	0.33	0.58	372	2	0.84	6	UB$_1$
$n = 1000$	0.23	0.02	0.56	218	1	0.87	6	UB$_1$

To perform parameter tuning we made use of the automatic configuration tool *irace* [12]. For benchmark set RANDOM, one tuning instance for every combination of m and $|\Sigma|$ was used. Similarly, for benchmark set NON-RANDOM, one instance for every combination of m and *type* was used. This means 15 tuning instances were used for each of the parameter tuning runs concerning the RANDOM benchmark set, while 10 instances were used for each NON-RANDOM one. Every tuning was allowed a budget of 5000 algorithm runs.

4.1 Benchmark Set RANDOM

The results obtained for benchmark set RANDOM are reported in Tables 3, 4 and 5. These contain the results for the instances with $n = 100$, $n = 500$ and $n = 1000$ respectively. For each combination of n, m and $|\Sigma|$ and for each algorithm we present the average length of the best solutions found ($\overline{|s|}$) and the average time required for finding these best solutions ($\overline{t}_{best}[s]$). As each group consists of ten instances, and each algorithm was applied ten times to each instance, the results for each of the 45 table rows average over 100 runs. In each row, the best result is shown in bold.

We can observe that results for this benchmark set differ not by much among the two solution approaches, although BRKGA-LEARN performs more often slightly better. Note that, for this set of instances, the original BRKGA used the first greedy information design which simply consists of biasing cuts towards the middle of every string.

As these instances consist of uniform strings, this approach already produces a good prediction on the optimal cut point, which did not leave our proposed guidance much room for improvement.

Table 3. Comparison of BRKGA-LEARN and BRKGA on RANDOM instances with string length $n = 100$.

m	$|\Sigma|$	BRKGA-LEARN		BRKGA	
		$\overline{|s|}$	$\overline{t}_{best}[s]$	$\overline{s}$	$\overline{t}_{best}[s]$
10	4	**28.44**	98.36	28.34	19.06
10	12	**8.94**	11.73	8.00	0.06
10	20	**4.20**	0.04	4.00	0.00
50	4	19.02	144.95	**19.94**	85.65
50	12	**4.00**	0.03	4.00	1.25
50	20	**1.40**	0.39	0.20	0.00
100	4	15.72	168.17	**17.16**	68.45
100	12	**2.42**	30.20	2.20	5.38
100	20	**0.12**	12.91	0.00	0.00
150	4	13.16	160.99	**15.94**	53.37
150	12	**2.00**	0.02	2.00	0.20
150	20	**0.00**	0.00	**0.00**	0.00
200	4	12.02	34.86	**14.78**	92.28
200	12	**2.00**	0.29	1.60	1.08
200	20	**0.00**	0.00	**0.00**	0.00

Table 4. Comparison of BRKGA-LEARN and BRKGA on RANDOM Instances With string length $n = 500$.

m	$|\Sigma|$	BRKGA-LEARN		BRKGA	
		$\overline{|s|}$	$\overline{t}_{best}[s]$	$\overline{|s|}$	$\overline{t}_{best}[s]$
10	4	**159.56**	235.68	158.94	226.69
10	12	**59.90**	47.88	59.60	51.22
10	20	**36.12**	11.69	36.04	7.97
50	4	**126.40**	297.95	125.76	146.66
50	12	**40.06**	65.44	40.00	52.29
50	20	**22.00**	25.71	21.82	14.19
100	4	**117.26**	246.59	116.82	102.85
100	12	**34.30**	55.78	34.12	21.53
100	20	**18.00**	0.70	**18.00**	0.60
150	4	**112.50**	125.87	**112.50**	64.91
150	12	**32.00**	2.71	**32.00**	2.14
150	20	**16.02**	0.99	16.00	0.16
200	4	109.96	109.96	**110.00**	26.02
200	12	**30.08**	17.80	30.04	6.96
200	20	15.76	74.79	**15.90**	73.57

Table 5. Comparison of BRKGA-LEARN and BRKGA on RANDOM instances with string length $n = 1000$.

m	$|\Sigma|$	BRKGA-LEARN		BRKGA	
		$\overline{|s|}$	$\overline{t}_{best}[s]$	$\overline{|s|}$	$\overline{t}_{best}[s]$
10	4	**324.42**	155.60	323.98	164.38
10	12	125.44	78.44	**125.78**	110.02
10	20	77.68	46.47	**78.02**	107.96
50	4	**263.90**	88.78	263.74	96.49
50	12	**87.62**	99.69	**87.62**	87.74
50	20	50.16	17.68	**50.54**	43.04
100	4	**249.30**	70.03	248.84	78.81
100	12	78.36	44.55	**78.48**	58.38
100	20	**44.00**	3.68	**44.00**	1.76
150	4	**242.22**	62.90	242.06	73.12
150	12	74.26	40.61	**74.42**	49.93
150	20	41.26	100.94	**41.32**	87.43
200	4	**237.50**	64.25	237.32	69.46
200	12	**72.02**	19.07	72.00	13.19
200	20	39.88	124.41	**39.94**	77.83

4.2   Benchmark Set NON-RANDOM

The results for benchmark set NON-RANDOM are shown in Tables 6, 7 and 8, which follow the same structure as outlined in the last section.

Table 6. Comparison of BRKGA-LEARN and BRKGA on NON-RANDOM instances with string length $n = 100$.

m	$type$	BRKGA-LEARN		BRKGA	
		$\overline{|s|}$	$\overline{t}_{best}[s]$	$\overline{|s|}$	$\overline{t}_{best}[s]$
10	1	**32.24**	97.06	32.14	114.59
10	2	30.56	66.05	**30.84**	95.25
50	1	**25.78**	151.84	24.98	227.49
50	2	**25.28**	100.28	24.86	183.25
100	1	**22.16**	107.93	19.34	180.61
100	2	**21.98**	120.53	20.08	149.02
150	1	**19.36**	127.27	16.76	154.21
150	2	**19.76**	161.05	16.68	147.78
200	1	**18.10**	136.83	14.72	145.22
200	2	**18.58**	120.13	14.70	160.61

Table 7. Comparison of BRKGA-LEARN and BRKGA on NON-RANDOM instances with string length $n = 500$.

m	$type$	BRKGA-LEARN		BRKGA	
		$\overline{|s|}$	$\overline{t}_{best}[s]$	$\overline{|s|}$	$\overline{t}_{best}[s]$
10	1	66.18	39.68	**70.92**	129.82
10	2	**70.14**	60.28	64.78	117.03
50	1	58.58	160.06	**59.16**	287.05
50	2	**60.76**	218.02	55.32	187.35
100	1	**52.58**	205.24	49.30	321.75
100	2	**53.60**	222.17	51.00	277.44
150	1	**51.08**	286.49	46.52	344.74
150	2	**48.80**	285.49	48.78	315.31
200	1	**48.18**	327.75	43.62	354.31
200	2	**45.60**	368.48	43.60	407.89

Table 8. Comparison of BRKGA-LEARN and BRKGA on NON-RANDOM instances with string length $n = 1000$.

m	$type$	BRKGA-LEARN		BRKGA	
		$\overline{|s|}$	$\overline{t}_{best}[s]$	$\overline{|s|}$	$\overline{t}_{best}[s]$
10	1	90.50	96.65	**91.14**	113.29
10	2	90.92	130.95	**91.38**	102.84
50	1	**66.16**	154.49	65.40	284.90
50	2	**66.28**	157.41	63.94	255.81
100	1	**61.66**	228.93	59.60	346.65
100	2	**60.32**	223.04	57.88	266.50
150	1	**57.70**	286.83	55.74	306.60
150	2	**57.20**	258.09	54.38	195.92
200	1	**54.62**	282.27	52.46	72.89
200	2	**54.48**	268.08	52.06	149.85

In this case, BRKGA-LEARN obtains consistently and significantly better results than the original BRKGA with the exception of the instances with a very low number of input strings (m), for which the results are inconclusive. For all other instances, BRKGA-LEARN obtains better solutions than BRKGA, indicating a clear benefit from the proposed ML guidance in the case of non-uniform strings.

In order to measure the statistical significance of the differences between the obtained solution lengths of BRKGA-LEARN and BRKGA on the NON-RANDOM benchmark set, we employed the signed-rank Wilcoxon test [19]. It tests the one-sided alternative hypothesis that a solution value obtained by BRKGA-LEARN is in the expected case larger than the corresponding solution value obtained by BRKGA. We obtained a p-value of less than 10^{-4} indicating that the differences observed are statistically highly significant.

5   Conclusions and Future Work

This paper presented an example of how machine learning can be used to improve the performance of genetic algorithms. Particularly, we introduced a neural network based guidance for a biased random key genetic algorithm (BRKGA) applied to the longest common square subsequence (LCSqS) problem. BRKGA works on the basis of reducing from the LCSqS problem to the well-known longest common subsequence (LCS) problem. This is done by cutting each input string into two parts. The BRKGA is used to explore the set of possible cut points in the search for the best possible cut points. The presented machine learning guidance is implemented in order to leverage this search process. Given a string from a set of input strings, individual features for each input string are extracted, together with global ones. These are then fed into a neural network whose task it is to provide a presumably good cut point for the string. This information is then used inside BRKGA as greedy information.

We have experimentally evaluated the enhanced BRKGA (BRKGA-LEARN) against the original BRKGA using two sets of benchmark instances from the literature. BRKGA-LEARN has significantly improved the results of BRKGA for non-uniform instances. On the other hand, for the random instances, BRKGA turned out to not benefit as much from the machine learning guidance. The reason behind is that the original BRKGA naturally preferred greedy value for this benchmark set that simply consisted in biasing cut points towards the middle of the input strings.

As for future work, concerning the methodological aspects it would be interesting to try other machine learning models for regression as a replacement for the neural net and to extend the set of used features for the input strings. Concerning the training, another promising approach would be the application of reinforcement learning. Last but not least, the principles of the proposed learning-based approach are rather general, and it appears promising to apply this learning-based framework also to other BRKGA approaches for different hard optimization problems.

References

1. Bebis, G., Georgiopoulos, M.: Feed-forward neural networks. IEEE Potentials **13**(4), 27–31 (1994)
2. Bengio, Y., Lodi, A., Prouvost, A.: Machine learning for combinatorial optimization: a methodological tour d'horizon. Eur. J. Oper. Res. **290**(2), 405–421 (2021)
3. Blum, C., Blesa, M.J., Lopez-Ibanez, M.: Beam search for the longest common subsequence problem. Comput. Oper. Res. **36**(12), 3178–3186 (2009)
4. Dagum, L., Menon, R.: OpenMP: an industry standard API for shared-memory programming. IEEE Comput. Sci. Eng. **5**(1), 46–55 (1998)
5. David, O.E., Greental, I.: Genetic algorithms for evolving deep neural networks. In: Proceedings of the Companion Publication of the 2014 Annual Conference on Genetic and Evolutionary Computation, pp. 1451–1452 (2014)
6. Ding, S., Su, C., Yu, J.: An optimizing BP neural network algorithm based on genetic algorithm. Artif. Intell. Rev. **36**, 153–162 (2011)

7. Djukanovic, M., Raidl, G.R., Blum, C.: A beam search for the longest common subsequence problem guided by a novel approximate expected length calculation. In: Nicosia, G., Pardalos, P., Umeton, R., Giuffrida, G., Sciacca, V. (eds.) LOD 2019. LNCS, vol. 11943, pp. 154–167. Springer, Cham (2019). https://doi.org/10.1007/978-3-030-37599-7_14

8. Djukanovic, M., Raidl, G.R., Blum, C.: A heuristic approach for solving the longest common square subsequence problem. In: Moreno-Diaz, R., Pichler, F., Quesada-Arencibia, A. (eds.) Computer Aided Systems Theory – EUROCAST 2019. EUROCAST 2019. LNCS, vol. 12013, pp. 429–437. Springer, Cham (2020). https://doi.org/10.1007/978-3-030-45093-9_52

9. Inoue, T., Inenaga, S., Hyyrö, H., Bannai, H., Takeda, M.: Computing longest common square subsequences. In: 29th Annual Symposium on Combinatorial Pattern Matching, CPM 2018. Schloss Dagstuhl-Leibniz-Zentrum für Informatik, Dagstuhl Publishing (2018)

10. Kingma, D.P., Ba, J.: Adam: a method for stochastic optimization. arXiv preprint arXiv:1412.6980 (2014)

11. Luce, G., Frédéric Myoupo, J.: Application-specific array processors for the longest common subsequence problem of three sequences. Parallel Algorithms Appl. **13**(1), 27–52 (1998)

12. López-Ibáñez, M., Dubois-Lacoste, J., Stützle, T., Birattari, M.: The irace package: iterated racing for automatic algorithm configuration. Oper. Res. Perspect. **3** (2011)

13. Maier, D.: The complexity of some problems on subsequences and supersequences. J. ACM **25**(2), 322–336 (1978)

14. Nakatsu, N., Kambayashi, Y., Yajima, S.: A longest common subsequence algorithm suitable for similar text strings. Acta Informatica **18**, 171–179 (1982)

15. Rahim Khan, M.A., Zakarya, M.: Longest common subsequence based algorithm for measuring similarity between time series: a new approach. World Appl. Sci. J. **24**(9), 1192–1198 (2013)

16. Reixach, J., Blum, C., Djukanovic, M., Raidl, G.: A biased random key genetic algorithm for solving the longest common square subsequence problem. SSRN: https://ssrn.com/abstract=4504431 or https://doi.org/10.2139/ssrn.4504431 (2023)

17. Wang, Q., Pan, M., Shang, Y., Korkin, D.: A fast heuristic search algorithm for finding the longest common subsequence of multiple strings. In: Proceedings of the AAAI Conference on Artificial Intelligence, vol. 24, pp. 1287–1292 (2010)

18. Wang, Y.: Longest common subsequence algorithms and applications in determining transposable genes. arXiv preprint arXiv:2301.03827 (2023)

19. Woolson, R.F.: Wilcoxon signed-rank test. Wiley Encyclopedia of Clinical Trials, pp. 1–3 (2007)

Sparse Surrogate Model for Optimization: Example of the Bus Stops Spacing Problem

Valentin Vendi⬤, Sébastien Verel$^{(\boxtimes)}$⬤, and Cyril Fonlupt$^{(\boxtimes)}$⬤

Univ. Littoral Côte d'Opale, LISIC, 62100 Calais, France
{valentin.vendi,verel,cyril.fonlupt}@univ-littoral.fr

Abstract. Combinatorial optimization problems can involve computationaly expensive fitness function, making their resolution challenging. Surrogate models are one of the effective techniques used to solve such black-box problems by guiding the search towards potentially good solutions. In this paper, we focus on the use of surrogate based on multinomial approaches, particularly based on Walsh functions, to tackle pseudo-Boolean problems. Although this approach can be effective, a potential drawback is the growth of the polynomial expansion with problem dimension. We introduce a method for analyzing real-world combinatorial black-box problems defined through numerical simulation. This method combines Walsh spectral analysis and polynomial regression. Consequently, we propose a sparse surrogate model that incorporates selected, relevant terms and is simpler to optimize. To demonstrate our approach, we apply it to the bus stop spacing problem, an exemplary combinatorial pseudo-Boolean challenge.

Keywords: Surrogate · Sparse model · Mobility problem · Walsh basis

1 Introduction

In industry 4.0 or in academic research such as in chemistry science, ocean science, energy system or transportation system [30], digital twins have evolved into tools for modeling systems and analyzing them through numerical simulations. This evolution gave birth to Simulation-Based Optimization (SBO) [16] which aims to solve optimization problems based on numerical simulation. However, SBO leads to additional challenges. Mainly, evaluating a candidate solution may be highly time-consuming ranging from a few seconds to hours [2,8]. Furthermore, the optimization problems are often considered as black-box problems where no algebraic definition is available. To address this problem, three main strategies can be used. Parallel approaches benefit from larger computational resources to reduce evaluation time. As the number of evaluated candidate solutions is limited, researchers also define optimization algorithms to increase the convergence rate toward the most promising solutions [24]. Lastly, Surrogate-Assisted Optimization (SAO) builds an algebraic model from evaluated solutions,

© The Author(s), under exclusive license to Springer Nature Switzerland AG 2024
T. Stützle and M. Wagner (Eds.): EvoCOP 2024, LNCS 14632, pp. 16–32, 2024.
https://doi.org/10.1007/978-3-031-57712-3_2

substituting the original optimization function with the surrogate to guide the search.

Although most research dealing with surrogate focus on numerical optimization [7], this article emphasizes on surrogate model in the case of pseudo-boolean optimization. Several surrogate models (see Sect. 2.1) such as neural network [15], random forest [17], kernel-based methods [42] and multinomial approaches are used, the latter being the most efficient [25]. The multinomial approaches decompose the original function into a multivariate polynomial. However, a significant limitation of multinomial approaches is the exponential increase in polynomial terms with order, which is a crucial parameter of the surrogate. Usually, this parameter is tuned by an expert based on the knowledge of interactions between variables. As the "shape", *i.e.* the algebraic structure of the model, of combinatorial real-world problems remains elusive, adding more knowledge into a multinomial surrogate model remains challenging even for an expert. Moreover, the primary goal of SAO is not to propose the most accurate machine learning model to approximate the original function, but to guide the search toward better candidate solutions. Thus, the surrogate model should be easy to optimize, and the optimal solution of the surrogate should guide the search to the optimal solutions of the original function.

The first objective of this paper is to propose a sparse surrogate model based on Fourier (Walsh) expansion, built with expert knowledge and easy to optimize using dynamic programming approach. The model is applied to the bus stop spacing problem, an optimization challenge involving selecting optimal bus stop positions in a city (see Sect. 2.3). This problem is defined using open source data, and the open source simulator MATSim for reproducibility. The second objective is to demonstrate how spectral Fourier/Walsh analysis of the real-world problem can be used to support hypothesis proposed by an expert in the application domain. To the best of our knowledge, this is the first spectral analysis of a real-world combinatorial optimization problem used to explore the algebraic properties of the fitness function. This work is a first step in turning a black-box optimization problem into a more transparent white/gray-box optimization problem [10].

The remainder paper is organized as follows. Section 2 reviews main works for combinatorial surrogate model, spectral analysis, and presents the bus stop spacing problem. Section 3 defines the sparse Walsh model. Section 4 presents the main experimental results on the spectral analysis, and the potential benefits of the proposed sparse model for optimization. The final section opens discussion, and perspectives of this work.

2   Related Work

This section outlines the main elements of our work. Starting from the definition of surrogate models and their application in combinatorial optimization, we then presents the use of spectral analysis in a context of explaining a surrogate Walsh model. Finally, we describe the bus stops spacing problem which serves as an illustration of our work.

2.1   Surrogate Model for Combinatorial Optimization

Surrogate-Assisted Optimization (SAO) uses a surrogate model to approximate the original, computationally expensive fitness function. In its most basic form, an offline version builds a surrogate model from an initial sample of solutions and optimizes an acquisition function to generate promising solutions. This acquisition function could be the surrogate function itself or a criterion such as expected improvement, etc. that is guiding the search by selecting a new candidate solution according to the surrogate model. The online SAO algorithm is based on the two same main components (learning a surrogate model and optimizing the acquisition function) but updates the surrogate model with generated solutions to refine it over iterations. (See Algorithm 1).

Algorithm 1. Surrogate-Assisted Algorithm.

$X \leftarrow$ initial sample
repeat
   $M \leftarrow$ Learn surrogate model of f from X
   $x^\star \leftarrow$ Optimize *w.r.t.* an acquisition function based on M
   $y^\star \leftarrow f(x^\star)$ using the numerical simulation
   $X \leftarrow X \cup \{(x^\star, y^\star)\}$
until stopping criterium

Surrogate models for combinatorial optimization have gained more and more attention [7] in recent years due to the progress of supervised machine learning techniques for learning heterogeneous data. Besides classical approaches from continuous optimization that replace the Euclidean distance by Hamming or other discrete distances in Krigging method [41] or kernel-based methods (Radial basis function) [27], advanced schemes use dedicated combinatorial structures. The BOCS method [6] uses multivariate polynomial of Boolean variables, estimated via Bayesian regression. This technique has been improved in COMBO [29] which uses a Cartesian product of graphs to represent discrete categorical variable in the framework of Bayesian optimization. More recently, to reduce the computation time of the parameters estimation of the multilinear polynomial model used for example in BOCS method, COMEX method [12] uses exponential weight updates from reinforcement learning. This approach, initially developed for pseudo-Boolean functions, has been extended to include categorical variables. [11].

Surrogate model based on Walsh functions [38], building multivariate polynomial models (see next Sect. 2.2), have proven to be effective in various contexts including interpolation where precise learning of Walsh coefficients is necessary [14], numerical simulations with some noise [25], and in multiobjective combinatorial optimization [13].

However, the surrogate part of BOCS, COMBO, or COMEX shows accuracy for combinatorial optimization for problem dimensions up to 50 and is usually optimized by a basic Simulated Annealing algorithm. In contrast, models based on Walsh functions take care of the optimization part and the properties of Walsh functions through efficient local search or evolutionary computation algorithms.

2.2 Spectral Analysis of Walsh Model

The Walsh basis, sometime also called Fourier basis, of pseudo-boolean functions space is an orthogonal, normal, and finite basis [28]. Any pseudo-boolean function $f : \{0,1\}^n \rightarrow \mathbb{R}$ can be uniquely represented as a linear combination of Walsh functions:

$$\forall x \in \{0,1\}^n, \ f(x) = \sum_{I \subset [n]} \beta_I \varphi_I(x) \text{ with } \varphi_I(x) = \prod_{i \in I} (-1)^{x_i}$$

$\beta_I \in \mathbb{R}$ is the coefficient associated to the Walsh function φ_I. Each Walsh function can be indexed by the subset of variables $I \subset [n]$. The order of a Walsh function φ_I is the size of the set I, *i.e.* the number of binary variables. As such, the Walsh expression can be written as a multilinear polynomial expression ranked by the order of terms from the constant term to the highest order which is the degree of the Walsh expansion:

$$f(\sigma) = \beta_\emptyset + \sum_{i \in [n]} \beta_i \sigma_i + \sum_{i<j \in [n]} \beta_{i,j} \sigma_i \sigma_j + \sum_{i<j<k \in [n]} \beta_{i,j,k} \sigma_i \sigma_j \sigma_k + \dots$$

Here $\sigma_i \in \{-1,1\}$ corresponds to the binary variable x_i such that $\sigma_i = (-1)^{x_i}$.

The constant term $\beta_\emptyset$ represents the average value of f across all solutions in $\{0,1\}^n$: $\mathbb{E}_x[f] = \beta_\emptyset$. The orthogonal and normal properties of the basis imply that the total variance of f is equal to the sum of square of its non-constant terms: $\mathrm{Var}_x[f] = \sum_{J \neq \emptyset} \beta_J^2$. As a consequence, each term can be easily understood. Each term $\beta_I \prod_{i \in I} \sigma_i$ indicates the interaction between the binary variables x_i for $i \in I$. The sign of β_I shows the sign of this interaction, while β_I^2 represents the part of f total variance of f explained by this interaction(*i.e.* the strength of interaction).

For instance, the linear terms β_i give the individual contribution of variables x_i to the fitness function, the terms $\beta_{i,j}$ is the quadratic contribution of variables x_i, and x_j, etc. From a geometric perspective, the variance of f is the square euclidean norm, and each β_I^2 is the squared length the projection onto the φ_I axis in the orthonormal Walsh basis. The square value β_I^2 is also called Fourier weight and defines the *spectral sample* of f (def. 1.17, and 1.18 [28]) as the probability distribution on $I \subset [n]$ is proportional to β_I^2. Then, we can define the weighted degree of f which is the average degree weighted by β_I^2: $\deg_\beta(f) = \frac{1}{\sum_{I \in [n]} \beta_I^2} \sum_{I \in [n]} \beta_I^2 |I|$.

With the goal of understanding the main interactions between variables which contributes to fitness function, we use a spectral analysis of pseudo-boolean functions [28]. Basically, we analyze Walsh coefficients normalized by the total variance:

$$\bar{\beta}_I = \frac{\beta_I}{\sqrt{\sum_{J \neq \emptyset} \beta_J^2}}$$

The normalized coefficients explain the importance of the corresponding interaction by the ratio of the total variance explained: $\bar{\beta}_I^2 = \frac{\beta_I^2}{\text{Var}_x[f]}$. The coefficient $\bar{\beta}_I^2$ is also called normalized amplitude spectrum [18]. Notice that the analysis of the amplitude spectrum has mainly been used to analyze the ruggedness of the fitness landscape [18,35]. Actually, the ruggedness which is linked to non-linearity of the fitness function can be measured by the autocorrelation coefficient, and can be deduced from the Walsh coefficients [34]. In this work, we propose to analyze directly the normalized spectrum. The strength of interaction can be compared across different functions, surrogate scenario, etc. and can be represented according to the order of terms or other special properties of the terms. The spectrum also helps to understand the "shape" of a real-world optimization function which also contribute to design better benchmarks of combinatorial optimization problems.

2.3 Bus Stops Spacing Problem

The bus stops spacing problem is a challenge in the field of transportation modeling [22,23,31,32,43,44]. It involves determining the optimal placement of bus stops in a public transportation system within a certain area. Many approaches to this problem exist in order to optimize different criteria of public transportation such as passengers travel time [23,43,44], economic costs [22,31], or environmental impact [32].

Various modeling approaches have been applied to this problem. Some utilize Thiessen polygons [43], or Voronoi diagrams [44] that use static properties of sub-divisions of the full road network for identifying optimal bus stop locations. Another approach uses a static model that incorporates multiple physical and economic constraints to compute the optimal bus stops spacing [31]. In [22], a bi-level optimization method aiming to minimize social costs of the transport system by finding the best positioning of bus stops evenly spaced along the road.

Agent-based simulators have emerged as another approach to design complex models of transportation systems. In [23] following [1], the authors use MATSim [21] as a simulation environment to compute bus passengers travel time according to their artificial scenario of mobility plan called SIALAC [25]. This method enhances the precision of the evaluation function, at the cost of an expensive computation time. In contrast to the milliseconds required for algebraic models, a single simulation of a scenario can take up to 1 min to compute.

New opportunities have emerged with the introduction of the Eqasim [19] pipeline that allows to create mobility scenarios for MATSim based on open-source publicly available data [20]. This pipeline has brought reproducible experiments in urban mobility context. It has been used to model various cities or regions such as Ile-de-France [20], Sao Paolo [3], or California [5]. The workflow of Eqasim from raw data to simulation is shown in Fig. 1. For the sake of reproducibility, all data and code of this article are available: https://gitlab.com/vvendi/offline-wsao. Given the accuracy of such model based on real-world data, we use this pipeline, and MATSim in the context of bus stops spacing problem resolution as described in next Sect. 4.

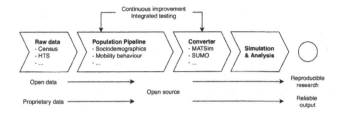

Fig. 1. Eqasim pipeline. Picture from the original article [19].

3   Sparse Walsh Model

In Surrogate Assisted Optimization (SAO), a surrogate model is learned from a sample of evaluated solutions in order to approximate the original fitness function and to guide the search towards better solutions. Thus, the surrogate model has to be frugal to estimate using a small sample of solutions, and easy to optimize in order to guide the search efficiently. One main drawback of multi-linear polynomial surrogate such as Walsh expansion is the increase in the number of terms with the degree. The number of terms of order k is $\binom{n}{k}$ and the number of terms of degree d is then: $\sum_{k=0}^{d} \binom{n}{k}$. For example, for a degree 3 Walsh expansion, the number of terms is $1 + n + \frac{n(n-1)}{2} + \frac{n(n-1)(n-2)}{6}$. Indeed, many works try to use sparse surrogate models with a low number of terms (see Sect. 2.1) in order to increase the accuracy of the surrogate given the small sample size. Here, in this work we also propose to use a sparse model, but on the contrary to previous works mainly based on data-guided method, we will use expert knowledge to select the most relevant terms of the surrogate model and consider the difficulty of optimization of the surrogate model.

The mean square error of a surrogate function $\hat{f}(x) = \sum_I \hat{\beta}_I \varphi_I(x)$ to approximate the fitness function $f(x) = \sum_I \beta_I \varphi_I(x)$ is given by the distance between functions:

$$\text{mse}(\hat{f}) = \mathbb{E}_x[(\hat{f}(x) - f(x))^2] = \sum_{I \subset [n]} (\hat{\beta}_I - \beta_I)^2$$

For instance, when the surrogate model is a truncation $\hat{f}_p$ of the original function f with p terms from a set $P \subset 2^{[n]}$ where: $\begin{cases} \hat{\beta}_I = \beta_I \ \forall I \in P \\ \hat{\beta}_I = 0 \ \ \forall I \notin P \end{cases}$. Then, the mean square error of the truncation is: $\text{mse}(\hat{f}_p) = \sum_{I \notin P} \beta_I^2$. As a consequence, the truncation with p terms minimizing the mean square error is composed by the p terms with the highest β_I^2-values. The design of a surrogate by an expert would push to select the most important variables interactions which influence the objective function. Heuristically based on knowledge of the optimization problems, some researchers propose to bound the degree of the Walsh expansion [6,13,14], but the number of terms can still be too large compare to the sample size. In feature selection machine learning problems [26], where the goal is to select the most important predictors for a machine learning algorithm, block model [4,33] is used which supposes that the features are clustered into families of variables. In this work, we go in this direction by hypothesizing that the order of variables in the binary string representation of a solution is not random, and gives useful information to exploit. We suppose that variables close in the representation impact the fitness function, and those far in the representation do no interact. Let's define the distance between variable x_i, and x_j in the binary string x as the distance between indexes: $|i - j|$, and the diameter of the set of variables I by: $D(I) = \max\{|i_1 - i_2| : (i_1, i_2) \in I^2\}$. The sparse surrogate model of degree d, and lag ℓ is defined by:

$$\hat{f}_{d,\ell}(x) = \sum_{\substack{I \subset [n] \\ \text{s.t. } |I| \leqslant d, D(I) \leqslant \ell}} \hat{\beta}_I \varphi_I(x)$$

The number of terms is reduced compared to a simple full expansion of degree d. For a degree d, we got $(n - \ell)\binom{d-1}{\ell} + \sum_{k=d-1}^{\ell-1} \binom{d-1}{k}$ terms. For example, for a problem dimension n, the sparse model of degree 2 with lag ℓ has $1 + n + (n - \ell)\ell + 1$ terms, which is linear with problem dimension, and not quadratic. Indeed, in evolutionary computation, this sparse model is known as k-bounded Walsh model [36], where each sub-function of the model depends only on k others variables.

Moreover, for k-bounded functions, a dynamic programming approach have been proposed to find in polynomial time the global minimum [39,40]. The same algorithm can be used for the sparse model of degree d and lag ℓ. First, a Walsh function f_n on $\{0,1\}^n$ can be split in two parts, the terms that do not contain the variable x_n, and the terms that contain x_n: $\forall x \in \{0,1\}^n$, $f_n(x) = f_{n-1}(x) + F_n(x)$ with $F_n = \sum_{I \subset [n] \text{ s.t. } n \in I} \beta_I \varphi_I(x)$. The lag ℓ ensures that F_n depends only on variables $x_{n-\ell}, \ldots, x_n$. So, the common variables between f_{n-1}, and F_n are the ℓ variables $x_{n-\ell}, \ldots, x_{n-1}$:

$$f_n(x_1, \ldots, x_{n-\ell}, \ldots, x_n) = f_{n-1}(x_1, \ldots, x_{n-\ell}, \ldots, x_{n-1}) + F_n(x_{n-\ell}, \ldots, x_{n-1}, x_n)$$

When those variables are fixed, the two parts f_{n-1}, and F_n are independent, and the global minimum of f_n is the sum of the minima:

$$
\min_{\substack{x \in \{0,1\}^n \text{ s.t. } (x_{n-\ell}, \ldots, x_{n-1}) = s}} f_n(x_1, \ldots, x_{n-\ell}, \ldots, x_{n-1}, x_n) =
$$

$$
\min_{\substack{x \in \{0,1\}^{n-1} \text{ s.t. } (x_{n-\ell}, \ldots, x_{n-1}) = s}} f_{n-1}(x_1, \ldots, x_{n-\ell}, \ldots, x_{n-1}) \quad + \quad \min_{\substack{x_n \in \{0,1\} \text{ s.t. } (x_{n-\ell}, \ldots, x_{n-1}) = s}} F_n(x_{n-\ell}, \ldots, x_{n-1}, x_n) \qquad (1)
$$

Defining the state S of the dynamic programming algorithm as the possible values of the variables $x_{n-\ell}, \ldots, x_{n-1}$, the Eq. 1 defines the recurrence formula to update the state of the algorithm. The last step of the algorithm selects the state value with the minimum fitness value. Notice that the state size is 2^ℓ, and the complexity is bounded by $\Theta(n 2^\ell)$ which is linear with problem dimension.

4   Experimental Analysis

4.1   Bus Stop Spacing Problem Design

In this section, we focus on an instance of the bus stops problem in the city of Calais, France. We make use of the MATSim simulator [21] coupled with the Eqasim pipeline [19] to run simulations from real-world data [20], which makes our simulation of the city of Calais very precise and reproducible, but with the drawback of a very expensive computation time, around 30 min per execution. We treat the bus stops spacing problem as a pseudo-boolean problem by modeling it as follows: All possible locations for bus stops are defined and represented as a binary variable representing the state of activation of the according bus stop. A solution is thus a binary string that represents the bus stops to activate in the simulation. In our case, we work on the bus line n°2 in Calais, which contains 76 bus stops, and we arbitrarily add 32 additional bus stops that could be potential candidates to improve the existing route for a total of 108 bus stops. Notice that this problem dimension is large in comparison with previous artificial benchmarks for surrogate-based combinatorial optimization [6,14]. Figure 2 shows the outward route and the return route of bus. A few of these stops perfectly fit with the original bus route, while others diverge, leading to detours. The purpose of these off-route stops is to introduce diverse options that our algorithm must identify and exclude from the final selection of bus stops. The 108 bus stops are ordered in a binary string solution such that the stop represented by the next bit is the next stop geographically. Our fitness fitness function defines a pseudo-Boolean problem: $f : \{0,1\}^n \to \mathbb{R}$ with a search space of binary strings of dimension $n = 108$, and the fitness function computes the average travel time of people using bus expressed in seconds. In practice, the 3-steps process to evaluate a candidate solution is the following:

- Edit relevant Eqasim input files, i.e. GTFS files which describe the public transportation system (schedule), according to the solution to evaluate
- Run Eqasim pipeline to generate MATSim input files
- Run MATSim simulation and get quality criterium i.e. average travel time of people using bus

4.2   Spectral Analysis

In this section, the Walsh/Fourier spectral analysis is computed to analyze (i) the most relevant degree of Walsh expansion which is the main parameter used in polynomial surrogate models (See Sect. 2.1), (ii) the relevance of the lag parameter of the proposed sparse model. The search space dimension of bus stop spacing problem is too large ($n = 108$) to compute exactly the Walsh coefficients. So, we compute the exact value of Walsh coefficients using full enumeration of solutions from a family of sub-spaces. In our work, we used sub-spaces composed by 8 contiguous bits in the binary string x which correspond to the 8 contiguous bus stops: from bit x_i to bit x_{i+7} for $i \in \{1, 5, 9, \ldots n - 7\}$. In each of these

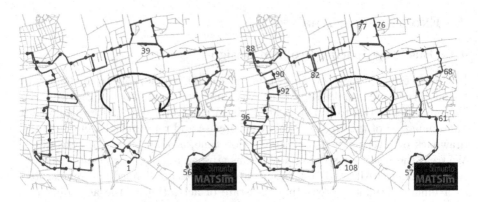

Fig. 2. Bus stops on the outward route and on the return route

sub-spaces, the bits that were not in the range of x_i to x_{i+7} were set to 1 *i.e.* the bus stop associated was turned on. Of course, this choice introduces bias in the estimation of Walsh coefficients, but it seems to be more meaningful (instead of a random binary value) for the bus stop problem. A same Walsh coefficient may appear in several sub-spaces. To estimate the weights, we compute the average of the normalized Fourier weight $\bar{\beta}_I^2$ across all sub-spaces which contains the coefficient $I \subset [n]$.

Figure 3 (left side) shows the distribution of the normalized Fourier weights $\bar{\beta}_I^2$ according to the order. The values are heterogeneous, so notice the log-scale of values. The terms of order 1, and 2 are the most important terms in the expansion. The median of order 1 weights is 17 times larger than order 2, and the median of order 2 weights is 1.7 times larger than order 3. From order 4 to 8, the ratio is much smaller. The sum of normalized Fourier weights of order 1 is 0.544, so 54.4% of the variance is explained by order 1 coefficients. Similarly, 79% of the variance is explained by order 1 and 2 combined. The weighted degree of the Walsh expansion (See definition in Sect. 2.2) is 1.66 between the order 1 and 2. Those first results suggest that a degree 2 Walsh expansion should bring an relevant surrogate model.

First, we analyze more precisely the linear impact of binary variables. Figure 4 shows the distribution of the logarithm of the normalized Fourier weights $\bar{\beta}_I^2$ of order 1. The distribution shapes an unimodal distribution. The median is around 0.7×10^{-3}, and the standard deviation of the log-values is 0.877. The distribution shows the Fourier weights are highly heterogeneous. Some linear contributions of bus stop are negligible, and at the opposite, some bus stops highly impact the fitness function. The 10 most important variables in decreasing rank of importance have identifier 96, 88, 82, 92, 90, 39, 76, 61, 77 and 68. Most of them match some extrapolated bus stops on the map that we can see in the Fig. 2. Indeed, the Fourier/Walsh analysis is easily explainable on the map for an expert in mobility transportation. Focusing on Walsh coefficients of order 2, Fig. 3 (right side) shows the distribution using boxplot of normalized

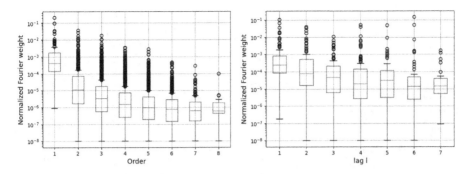

Fig. 3. Normalized Fourier weight. Left: as a function of order. Right: only for order 2 terms as a function of lag. Notice the y-log scale.

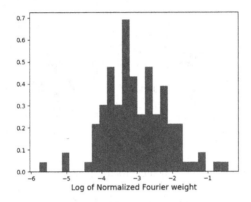

Fig. 4. Distribution of the logarithm of normalized Fourier weights $\bar{\beta}_I^2$ of order 1.

Fourier weights of order 2 coefficients according to the lag parameter which is the distance $|i - j|$ in binary string between variables i, and j. The weights importance decreases with the lag. The median of the lag 1 weights is 3.24 times larger than the median of lag 2, and the lag 2 median is 1.67 times larger than the median of lag 3 coefficients. From lag 4 to 8, the ratio of medians is much smaller. Coefficients with lag 1 and 2 explain 51.4% of the order 2 coefficients variance. Indeed, some very specific coefficients with lag 6 impact the fitness function as detailed in the next paragraph.

Figure 5 shows the Fourier spectrum of order 2 terms for 4 specific sub-spaces with variables $(x_i, \ldots, x_{i+7})$. The normalized Fourier weight $\bar{\beta}_{i,j}^2$ of term for variables i, j is displays in matrix at position (i, j). Thus, the matrix is symmetric, and the values on the diagonal are set to zero. All sub-spaces can not be displayed, so 4 representatives examples are selected. The one on the top-left shows a generic case, the highest weights are close to the diagonal *i.e.* are close to each others, with the highest weight representing the interaction between two stops at distance 3. The top-right with sub-space $(17, \ldots, 24)$ is another example very similar to the top-left one. Once again, the highest weights

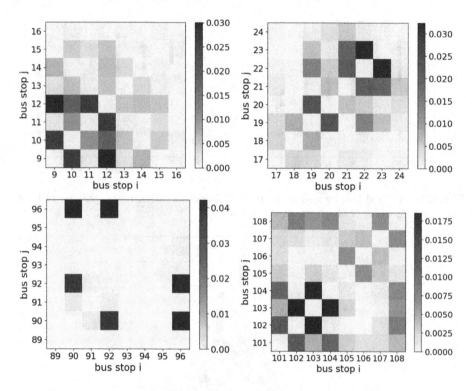

Fig. 5. Fourier spectrum of order 2 coefficients for 4 sub-spaces. The matrix value (i, j) displays the normalized Fourier weight $\bar{\beta}^2_{i,j}$ of the quadratic term which corresponds to bus stops with ids i, and j. Color is the intensity of $\bar{\beta}^2_{i,j}$-values.

are close to the diagonal with the maximum weight between the two stops 22 and 23 at distance 1. The sub-space with variables $(89, \ldots, 96)$ at the bottom-left is a specific case where the terms with the highest weights are not the ones with the smallest lag distance near the diagonal. Indeed, bus stops 90, 92 and 96 are among those with the highest linear importance and also show quadratic importance even though the lag is equal to 6. These bus stops are displayed on Fig. 2 on the return route. Intuitively, we can make the hypothesis that this combination of "side-roads" would have an impact on the travel time, which is confirmed with the Fourier analysis of the fitness function. The sub-space with variables $(101, \ldots, 108)$ at the bottom-right is an example where all weights are pretty low leading to an uniform distribution of weights, but the most important quadratic terms are still close to the diagonal. The spectral analysis explains the degree of a relevant Walsh expansion, the main interaction between variables, and highlight the sparsity of an efficient surrogate model that can be used. The next step is to compare the candidate sparse models with the previous proposed models in the literature.

4.3   Sparse Walsh Model Quality

In this section, the quality of the sparse Walsh surrogate models is compared to the quality of the full Walsh surrogate [25]. We follow the sparse regression method based on the classical LASSO-LARS algorithm from [13,23,25] to estimate the parameters of the Walsh expansion which contains all terms of order below a given degree k, so called full Walsh expansion of degree k in this article. Figure 6 (left) compares the R^2 coefficient of determination (part of explained variance) of the different surrogate models: full Walsh expansion of degree 1, and of degree 2, and the sparse Walsh model with lag $\ell = 4$ of degree 2, and of degree 3. The R^2 coefficient is estimated on an independent random test set of size 400, and the surrogate is trained on random sample size up to 3,600 solutions. As in previous studies [38], for small sample size, linear Walsh expansion with only $n + 1$ terms is more accurate than quadratic Walsh expansion with $1 + n + n(n - 1)/2$ terms, but becomes more accurate for larger sample size of 2,000 solutions. However, sparse Walsh model of degree 2 outperforms full Walsh expansion of degree 1 and 2 for almost any training sample size, but also the sparse model of degree 3 until the training set reach a size of 2700 solutions which confirms the previous analysis of Sect. 4.2 on the importance of low order terms. Remember that computation time to evaluate a single solution is about 30 min making the maximum training sample size very large compared to typical computation effort of an optimization algorithm (which is often $10n$), in this regard, we can say that the sparse model of order 2 is better than the other tested models. Figure 6 (right) compares the R^2 regression quality of sparse model of degree 2 for different lag parameter values. For small train sample until size 700, all sparse models have approximately the same quality. Between sample size 700, and 2,200, the sparse with lag 2 slightly outperforms the other ones, and for the largest sample size, sparse model with lag 1 is outperformed by all the other ones. Overall, the sparse model of degree 2 with lag $\ell = 2$ seems to be efficient according to R^2 quality.

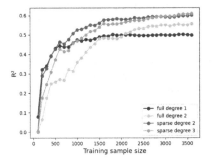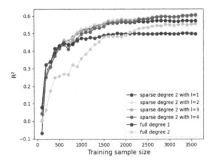

Fig. 6. R^2 estimated on a test set according to train sample size for different surrogate models. Left: comparison of degrees with sparse models of lag $\ell = 4$. Right: comparison of different lags with degree 2 models.

4.4   Optimization with the Sparse Surrogate Model

The sparse Walsh model is designed to incorporate expert knowledge in the surrogate model for which the relevance is supported by the previous analysis, but also to be efficient to optimize. Indeed, in Sect. 3, we show that the optimal solution of the sparse surrogate can be found in linear time with problem dimension. In addition to the regression quality of the sparse surrogate, in this section we show the ability to find solutions with low fitness value using sparse surrogate model for the bus stop spacing problem instance.

In this preliminary study, we do not follow the anytime surrogate-assisted Algorithm 1 and only use the offline version. Only the original fitness quality of the solutions which are the minimum of the surrogate function is analyzed. We compared two different surrogates models: the first one is a full Walsh model of degree 2, and the other one is a sparse Walsh model of degree 2 with lag $\ell = 2$ (see Sect. 4.3). For each surrogate model, we learn 30 surrogate models trained on independent random samples of medium size $5n = 540$, large size $10n = 1080$, and very large size of 3000 solutions. Medium, and large size are typical sizes used in the literature [6,12]. Random solutions are sampled from the original random sample of 4,000 solutions by random sub-sampling technique. To optimize each surrogate models, we used the efficient iterated local search DRILS that uses partition cross-over, and the variable interaction graph perturbation[1] [10,37] for the full Walsh surrogate models; and dynamic programming (Sect. 2.1) for the sparse Walsh surrogate models. Table 1 shows the average fitness obtained, and the Mann-Whitney statistical test at the level of 5%. As a comparison, the average fitness of the 4,000 random solutions is 2,746 with a standard deviation of 13.1 and a minimum of 2,712. The fitness of the existing bus stops in Calais is 2,698. The sparse model with dynamic programming outperformed the full Walsh model with DRILS for medium, and large sizes with a larger difference for the medium size. For very large size, both methods have similar performances, however notice that the variance of sparse models is always smaller which suggests a more robust optimization process. The sparse model of degree 2 and lag $\ell = 2$ is able to guide the search toward promising good solutions, and seems to converge quicker $w.r.t.$ to sample size than the full model of degree 2.

Table 1. Average, and standard deviation of original fitness obtained from the optimization of different surrogate models for different size of training sample size $|X|$. Bold highlights statistical difference according to Mann-Whitney test with level of 5%.

$|X|$	Full Walsh model	Sparse Walsh model
540	2,715 (7.5)	**2,697** (4.8)
1,080	2,704 (4.5)	**2,694** (3.6)
3,000	**2,692** (3.7)	**2,691** (1.9)

---

[1] Recommended parameters values, and stopping criterion at 2 s of computation.

5   Conclusions, and Perspectives

In this paper, inspired by block-model, we propose a sparse Walsh surrogate model which incorporates expert knowledge based on the hypothesis that the representation, *i.e.* the encoding, of solution is not random: close variables in the encoding could interact more than the other ones. Moreover, the sparse model is dedicated for efficient optimization. We propose a Walsh/Fourier spectral analysis of the fitness function of a real-world problem. This analysis shows that it is possible to use surrogate models based on Walsh expansion to help the expert to understand the real-world problem in addition to good optimal solutions given by the optimization process. As such, this work is a step forward to explainable optimization to bring decision supports around an optimization problem.

This initial works including the spectral analysis could be extended to other black-box combinatorial problems either to problems with binary representations, or to more complex representation such as permutations space [9]. In this work, as a first step for the optimization process, we use an offline optimization scenario. Anytime surrogate-assisted optimization which updates the sample of solutions during the optimization process has to be tested. Obviously, we plan to deal with other public transportation plans for the city of Calais with other criteria of interest, but also for larger cities, where the number of decision variables is much larger than the state-of-the-art surrogate-assisted algorithms, and requires parallel optimization algorithms.

Acknowledgments. Experiments presented in this paper were carried out using the CALCULCO computing platform, supported by DSI/ULCO. We thank Sébastian Hörl for his help in setting up the Eqasim environment and the modification he made for our research project.

References

1. Armas, R., Aguirre, H., Tanaka, K.: Multi-objective optimization of level of service in urban transportation. In: Proceedings of the Genetic and Evolutionary Computation Conference, pp. 1057–1064 (2017)
2. Armas, R., Aguirre, H., Zapotecas-Martínez, S., Tanaka, K.: Traffic signal optimization: minimizing travel time and fuel consumption. In: Bonnevay, S., Legrand, P., Monmarché, N., Lutton, E., Schoenauer, M. (eds.) EA 2015. LNCS, vol. 9554, pp. 29–43. Springer, Cham (2016). https://doi.org/10.1007/978-3-319-31471-6_3
3. Aurore Sallard, M.B., Hörl, S.: An open data-driven approach for travel demand synthesis: an application to são paulo. Reg. Stud. Reg. Sci. 8(1), 371–386 (2021). https://doi.org/10.1080/21681376.2021.1968941
4. Bai, Z., Nguyen, H., Davidson, I.: Block model guided unsupervised feature selection. In: Proceedings of the 26th ACM SIGKDD International Conference on Knowledge Discovery & Data Mining, pp. 1201–1211 (2020)
5. Balac, M., Hörl, S.: Synthetic population for the state of California based on open data: examples of the San Francisco bay area and San Diego county, February 2021

6. Baptista, R., Poloczek, M.: Bayesian optimization of combinatorial structures. In: International Conference on Machine Learning, pp. 462–471. PMLR (2018)
7. Bartz-Beielstein, T., Zaefferer, M.: Model-based methods for continuous and discrete global optimization. Appl. Soft Comput. **55**, 154–167 (2017)
8. Branke, J.: Simulation optimisation: tutorial. In: Proceedings of the Genetic and Evolutionary Computation Conference Companion, pp. 862–889 (2019)
9. Chicano, F., Derbel, B., Verel, S.: Fourier transform-based surrogates for permutation problems. In: Proceedings of the Genetic and Evolutionary Computation Conference, pp. 275–283 (2023)
10. Chicano, F., Whitley, D., Ochoa, G., Tinós, R.: Optimizing one million variable NK landscapes by hybridizing deterministic recombination and local search. In: Proceedings of the Genetic and Evolutionary Computation Conference, pp. 753–760 (2017)
11. Dadkhahi, H., Rios, J., Shanmugam, K., Das, P.: Fourier representations for blackbox optimization over categorical variables. In: Proceedings of the AAAI Conference on Artificial Intelligence, vol. 36, pp. 10156–10165 (2022)
12. Dadkhahi, H., et al.: Combinatorial black-box optimization with expert advice. In: Proceedings of the 26th ACM SIGKDD International Conference on Knowledge Discovery & Data Mining, pp. 1918–1927 (2020)
13. Derbel, B., Pruvost, G., Liefooghe, A., Verel, S., Zhang, Q.: Walsh-based surrogate-assisted multi-objective combinatorial optimization: a fine-grained analysis for pseudo-Boolean functions. Appl. Soft Comput. **136**, 110061 (2023)
14. Dushatskiy, A., Alderliesten, T., Bosman, P.A.: A novel approach to designing surrogate-assisted genetic algorithms by combining efficient learning of Walsh coefficients and dependencies. ACM Trans. Evol. Learn. Optim. **1**(2), 1–23 (2021)
15. Dushatskiy, A., Mendrik, A.M., Alderliesten, T., Bosman, P.A.: Convolutional neural network surrogate-assisted GOMEA. In: Proceedings of the Genetic and Evolutionary Computation Conference, pp. 753–761 (2019)
16. Gosavi, A., et al.: Simulation-Based Optimization. Springer, New York (2015). https://doi.org/10.1007/978-1-4899-7491-4
17. Han, L., Wang, H.: A random forest assisted evolutionary algorithm using competitive neighborhood search for expensive constrained combinatorial optimization. Memetic Comput. **13**, 19–30 (2021)
18. Hordijk, W., Stadler, P.F.: Amplitude spectra of fitness landscapes. Adv. Complex Syst. **1**(01), 39–66 (1998)
19. Hörl, S., Balac, M.: Introducing the eqasim pipeline: from raw data to agent-based transport simulation. Procedia Comput. Sci. **184**, 712–719 (2021). the 12th International Conference on Ambient Systems, Networks and Technologies (ANT)/The 4th International Conference on Emerging Data and Industry 4.0 (EDI40) / Affiliated Workshops
20. Hörl, S., Balac, M.: Synthetic population and travel demand for Paris and île-de-France based on open and publicly available data. Transp. Res. Part C: Emerg. Technol. **130**, 103291 (2021)
21. Horni, A., Nagel, K., Axhausen, K. (eds.): Multi-Agent Transport Simulation MATSim. Ubiquity Press, London, August 2016
22. Ibeas, A., della Olio, L., Alonso, B., Sainz, O.: Optimizing bus stop spacing in urban areas. Transp. Res. Part E: Logist. Transp. Rev. **46**(3), 446–458 (2010)
23. Leprêtre, F., Fonlupt, C., Verel, S., Marion, V.: Combinatorial surrogate-assisted optimization for bus stops spacing problem. In: Biennial International Conference on Artificial Evolution (EA 2019). Mulhouse, France, October 2019

24. Leprêtre, F.: Fitness landscapes analysis and adaptive algorithms design for traffic lights optimization on SIALAC benchmark. Appl. Soft Comput. **85**, 105869 (2019)
25. Leprêtre, F., Verel, S., Fonlupt, C., Marion, V.: Walsh functions as surrogate model for pseudo-Boolean optimization problems. In: The Genetic and Evolutionary Computation Conference (GECCO 2019), pp. 303–311. Proceedings of the Genetic and Evolutionary Computation Conference, ACM, Prague, Czech Republic, July 2019
26. Li, J., et al.: Feature selection: a data perspective. ACM Comput. Surv. (CSUR) **50**(6), 94 (2018)
27. Moraglio, A., Kattan, A.: Geometric generalisation of surrogate model based optimisation to combinatorial spaces. In: Merz, P., Hao, J.-K. (eds.) EvoCOP 2011. LNCS, vol. 6622, pp. 142–154. Springer, Heidelberg (2011). https://doi.org/10. 1007/978-3-642-20364-0_13
28. O'Donnell, R.: Analysis of Boolean Functions. Cambridge University Press, Cambridge (2014)
29. Oh, C., Tomczak, J., Gavves, E., Welling, M.: Combo: combinatorial Bayesian optimization using graph representations. In: ICML Workshop on Learning and Reasoning with Graph-Structured Data (2019)
30. Pires, F., Cachada, A., Barbosa, J., Moreira, A.P., Leitão, P.: Digital twin in industry 4.0: technologies, applications and challenges. In: 2019 IEEE 17th International Conference on Industrial Informatics (INDIN), vol. 1, pp. 721–726. IEEE (2019)
31. Saka, A.A.: Model for determining optimum bus-stop spacing in urban areas. J. Transp. Eng. **127**(3), 195–199 (2001)
32. Saka, A.A.: Effect of bus-stop spacing on mobile emissions in urban areas (2003)
33. Saltiel, D., Benhamou, E.: Feature selection with optimal coordinate ascent (OCA). arXiv preprint arXiv:1811.12064 (2018)
34. Stadler, P.F.: Landscapes and their correlation functions. J. Math. Chem. **20**(1), 1–45 (1996)
35. Stadler, P.F.: Spectral landscape theory. Evolutionary dynamics: exploring the interplay of selection, accident, neutrality, and function, pp. 221–272 (2003)
36. Sutton, A.M., Whitley, L.D., Howe, A.E.: Computing the moments of k-bounded Pseudo-Boolean functions over hamming spheres of arbitrary radius in polynomial time. Theor. Comput. Sci. **425**, 58–74 (2012)
37. Tinós, R., Przewozniczek, M.W., Whitley, D.: Iterated local search with perturbation based on variables interaction for pseudo-Boolean optimization. In: Proceedings of the Genetic and Evolutionary Computation Conference, pp. 296–304 (2022)
38. Verel, S., Derbel, B., Liefooghe, A., Aguirre, H., Tanaka, K.: A surrogate model based on Walsh decomposition for Pseudo-Boolean functions. In: Auger, A., Fonseca, C.M., Lourenço, N., Machado, P., Paquete, L., Whitley, D. (eds.) PPSN 2018. LNCS, vol. 11102, pp. 181–193. Springer, Cham (2018). https://doi.org/10.1007/ 978-3-319-99259-4_15
39. Whitley, L.D., Chicano, F., Goldman, B.W.: Gray box optimization for MK landscapes (NK landscapes and MAX-kSAT). Evol. Comput. **24**(3), 491–519 (2016)
40. Wright, A.H., Thompson, R.K., Zhang, J.: The computational complexity of NK fitness functions. IEEE Trans. Evol. Comput. **4**(4), 373–379 (2000)
41. Zaefferer, M., Horn, D.: A first analysis of kernels for kriging-based optimization in hierarchical search spaces. In: Auger, A., Fonseca, C.M., Lourenço, N., Machado, P., Paquete, L., Whitley, D. (eds.) PPSN 2018. LNCS, vol. 11102, pp. 399–410. Springer, Cham (2018). https://doi.org/10.1007/978-3-319-99259-4_32

42. Zaefferer, M., Stork, J., Friese, M., Fischbach, A., Naujoks, B., Bartz-Beielstein, T.: Efficient global optimization for combinatorial problems. In: Proceedings of the 2014 Annual Conference on Genetic and Evolutionary Computation, pp. 871–878 (2014)
43. Zheng, C., Zheng, S., Ma, G.: The bus station spacing optimization based on game theory. Adv. Mech. Eng. **7**(2), 453979 (2015). https://doi.org/10.1155/2014/453979
44. Zhu, Z., Guo, X., Chen, H., Zeng, J., Wu, J.: Optimization of urban mini-bus stop spacing: a case study of Shanghai (China). Tehnicki Vjesnik **24**, 949–955 (2017)

Emergence of New Local Search Algorithms with Neuro-Evolution

Olivier Goudet[(✉)], Mohamed Salim Amri Sakhri, Adrien Goëffon, and Frédéric Saubion

Univ Angers, LERIA, SFR MATHSTIC, F-49000 Angers, France
{olivier.goudet,mohamedsalim.amrisakhri,
adrien.goeffon,frederic.saubion}@univ-angers.fr

Abstract. This paper explores a novel approach aimed at overcoming existing challenges in the realm of local search algorithms. The main objective is to better manage information within these algorithms, while retaining simplicity and generality in their core components. Our goal is to equip a neural network with the same information as the basic local search and, after a training phase, use the neural network as the fundamental move component within a straightforward local search process. To assess the efficiency of this approach, we develop an experimental setup centered around NK landscape problems, offering the flexibility to adjust problem size and ruggedness. This approach offers a promising avenue for the emergence of new local search algorithms and the improvement of their problem-solving capabilities for black-box problems.

Keywords: Neuro-evolution · Local search · Black-box optimization · NK landscapes

1 Introduction

Local search (LS) algorithms are commonly used to heuristically solve discrete optimization problems [10]. LS algorithms are usually composed of several components: a search space, a neighborhood relation, an evaluation function, and a selection strategy. The optimization problem instance to be solved can be fully defined by its set of feasible solutions —the decision space— and an objective function that must be optimized. A classic and direct use of local search, when applicable, is to consider the decision space as the search space, the objective function as the evaluation function, and a natural neighborhood relation, defined from an elementary transformation function (move) such as bitflip for pseudo-Boolean problems, or induced by specific operators for permutation problems [22].

Starting from an initial random solution, various components collaborate to drive the search towards optimal solutions. The effectiveness of this search process depends on the complexity of the problem, including factors such as deception and other structural characteristics [2,13,28,29]. The strategy to advance

© The Author(s), under exclusive license to Springer Nature Switzerland AG 2024
T. Stützle and M. Wagner (Eds.): EvoCOP 2024, LNCS 14632, pp. 33–48, 2024.
https://doi.org/10.1007/978-3-031-57712-3_3

in the search involves selecting neighboring solutions based on their evaluations, using a wide variety of criteria. These criteria can range from simple ones, such as choosing neighbors with better or the best evaluations, to more intricate approaches involving stochastic methods. The strategy is often derived from metaheuristic frameworks based on local search such as tabu search [6] or iterated local search [15]. The goal is to effectively leverage the available local and partial knowledge of the landscape to identify the most promising search paths that lead to optimal solutions.

Fitness landscape analysis [16] provides the optimization and evolutionary computation community with theoretical and practical tools to examine search landscapes. It allows for the assessment of problem characteristics and the evaluation of the performance of search algorithms. In the context of combinatorial fitness landscapes, these are represented as graphs defined by a discrete search space and a neighborhood relation. A fundamental challenge lies in developing a search algorithm capable of navigating a fitness landscape to reach the highest possible fitness value.

In general, achieving optimal solutions through a straightforward adaptive approach is quite challenging. This difficulty arises from the complex interplay among different parts of the solution, which can lead to locally optimal solutions. These local optima cannot be escaped by intensification or exploitative move strategies. Consequently, the optimization algorithm becomes stuck in a suboptimal state. The primary concern in such optimization processes is to strike a balance between exploiting promising search areas through greedy search methods and diversifying search trajectories by temporarily exploring less promising solutions.

To overcome this challenge, researchers have developed many metaheuristics [23] and even hyperheuristic schemes [20] to mix different strategies. These approaches typically incorporate parameters that allow for precise adjustment of the trade-off between exploration and exploitation. Still, they might not perform efficiently in a black-box context, as many heuristics leverage the unique properties of the given problem to solve it efficiently. Machine learning techniques have been widely used to improve combinatorial optimization solving [24] and to address the optimal configuration of solving algorithms. An approach to algorithm design known as "programming by optimization" (PbO) was introduced by Hoos [9]. This paradigm encourages algorithm developers to adopt and leverage extensive design possibilities that encompass a wide range of algorithmic techniques, to optimize performance for specific categories or groups of problem instances.

In various works, different machine learning approaches have been used either to consider offline adjustment, selection of parameters, or online control of the search process using reinforcement learning (RL) techniques (see [17] for a recent survey). The use of neural networks (NNs) in solving combinatorial optimization problems has been studied for decades [26], starting with the early work of Hopfield [11]. Recent applications of Graph Neural Networks in the context of combinatorial optimization have been proposed to reach optimal solutions or to assist the solving algorithm in proving the optimality of a given solution (see [4] for a recent survey).

In the traveling salesman problem, a GNN can be used to predict the regret associated with adding each edge to the solution to improve the computation of the fitness function of the LS algorithm [12]. Note that NNs are classically used in surrogate model-based optimization [30]. In [21], the authors introduce a GNN into a hybrid genetic search process to solve the vehicle routing problem. The GNN is used to predict the efficiency of search operators and to select them optimally in the solving process. A deep Q-learning approach has recently been proposed to manage the different stages of an LS-based metaheuristic to solve routing and job-shop problems [5]. High-level solving policies can often be managed by reinforcement learning in LS processes [27]. In this paper, our purpose is different, as we focus on building simple search heuristics for black-box problems, rather than scheduling specific operators or parameters. In particular, we assume that the learning process cannot be based on the immediate rewards that are used in RL. This motivates our choice of neuro-evolution.

Objective of the Paper. This study explores the potential for emerging search strategies to overcome existing challenges. The objective is to change how information is leveraged while retaining simple and generic search components. Considering a basic hill climber algorithm to achieve a baseline search process for solving black-box binary problem instances, our aim is to benefit from machine learning techniques to get new local search heuristics that will be built from basic search information instead of choosing a priori a predefined move heuristic (e.g., always select the best neighbor). Hence, our goal is to provide a NN with the same information as a basic LS and, after training, to use the NN as the basic move component of a simple LS process. To evaluate the efficiency of our approach, we define an experimental setup based on NK landscape problems, which allows us to describe a fitness landscape whose problem size and ruggedness (determining the number of local minima) can be adjusted as parameters.

2   General Framework

2.1   Pseudo-boolean Optimization Problems and Local Search

Let us consider a finite set $\mathcal{X} \subseteq \{0,1\}^N$ of solutions to a specific problem instance. These solutions are tuples of values that must satisfy certain constraints, which may or may not be explicitly provided. We evaluate the quality of these solutions using a pseudo-Boolean objective function $f_{\mathrm{obj}} : \mathcal{X} \to \mathbb{R}$. Therefore, a problem instance can be fully characterized by the pair $(\mathcal{X}, f_{\mathrm{obj}})$. In terms of solving this problem, $\mathcal{X}$ is referred to as the search space.

When solving an optimization problem instance with an LS algorithm, the objective is to identify a solution $x \in \mathcal{X}$ that optimizes the value of $f_{\mathrm{obj}}(x)$. Since we are primarily concerned with maximization problems, let us note that any minimization problem can be reformulated as a maximization problem. In this context, an optimal solution, denoted as $x^* \in \mathcal{X}$, must satisfy the condition that for every $x \in \mathcal{X}$, $f_{\text{obj}}(x) \leq f_{\text{obj}}(x^*)$. While exhaustive search methods, or branch and bound algorithms, guarantee the computation of an optimal solution, this is not the case with LS algorithms. However, computing optimal solutions within a reasonable time is often infeasible, leading to the use of local search algorithms within a limited budget of evaluations of f_{obj} to approximate near-optimal solutions.

LS algorithms operate within a structured search space, thanks to a fixed-sized neighborhood function denoted as $\mathcal{N} : \mathcal{X} \to 2^{\mathcal{X}}$. This function assigns a set of neighboring solutions $\mathcal{N}(x) \subseteq \mathcal{X}$ for each solution $x \in \mathcal{X}$. To maintain a fundamentally generic approach to LS, we assume that $\mathcal{N}$ is defined using basic flip functions, $flip_i : \mathcal{X} \to \mathcal{X}$, where $i \in [\![1, N]\!]$, and $flip_i(x)$ is equal to x except for the ith element, which is changed from 0 to 1 or vice versa. In this case, $\mathcal{N}(x) = \{flip_i(x) \mid i \in [\![1, N]\!]\}$. Starting from an initial solution, often selected randomly and denoted as x_0, LS constructs a path through the search space based on neighborhood relationships. This path is typically represented as a sequence of solutions bounded by a limit (horizon) H, denoted as $(x_0, x_1, \ldots, x_H)$, where for each $i \in [\![0, H-1]\!]$, $x_{i+1} \in \mathcal{N}(x_i)$. Let us denote $\mathcal{P} = \mathcal{X}^*$ the set of all paths (i.e., the set of all possible sequences built on $\mathcal{X}$).

In addition to the neighborhood function, this sequence of solutions is determined by a strategy, often involving the use of a fitness function f. For example, hill climbing algorithms select the next solution on the path based on a simple criterion: $\forall t \in [\![0, H-1]\!], f(x_t) < f(x_{t+1})$, where $f(x_t)$ represents the fitness evaluation of solution x_t. The process of choosing the next solution on the search path is referred to as a move. We denote $x_{t+1} = x_t \oplus flip_i$ the move that corresponds to $x_{t+1} = flip_i(x_t)$.

2.2 Local Search as an Episodic Tasks Process

According to previous remarks, an LS process can be modeled by a sequence of actions performed on states. To be as exhaustive as possible, these states must be general enough to describe the current solution, as well as the path already explored and future possible moves. Therefore, we may define a set of states as $\mathcal{S} = \mathcal{P} \times \mathcal{X} \times 2^{\mathcal{X}}$. Of course, states of $\mathcal{S}$ cannot be managed by a practical local search algorithm, at least from a memory size point of view. We introduce the notion of observation of a state as a function $o : \mathcal{S} \to \Omega$ where Ω is a domain that corresponds to an abstraction of the real states of search to gather only useful information for the considered local search strategy.

Note that here we are in the context of episodic (discrete states) tasks with a terminal state (the end of the search fixed for instance by a maximal number of moves H). Following a reinforcement learning-based description, a local search process can be encoded by a policy $\pi : \Omega \to \mathcal{A}$ where $\mathcal{A}$ is a set of actions. Here we consider deterministic policies, i.e. for each $o \in \Omega$, there exists one and only one action $a \in \mathcal{A}$, such that $\pi(o) = a$. Note that the policy can be parameterized by a parameter vector θ. The set of all policies is $\Pi = \{\pi_\theta \mid \theta \in \Theta\}$ where Θ is the parameter space.

In our context, we consider actions that are bitflips $flip_i$ defined in Sect. 2.1. Hence, $\mathcal{A} \supseteq \{flip_i \,|\, i \in [\![1, N]\!]\}$. Note that of course, other actions can be introduced to fit specific strategies.

We consider a transition function $\delta : \mathcal{X} \times \mathcal{A} \to \mathcal{X}$ such that $\delta(x, a) = x \oplus a$. The transition function turns an action that is defined on the set of observations to a move in the set of solutions. Note that the transition is used to update the current state and compute the next state of the LS process. An LS run can be fully characterized by the search trajectory that it has produced from an initial starting solution.

Given an instance $I = (\mathcal{X}, f_{\text{obj}})$, an initial solution $x_0 \in \mathcal{X}$, a policy π_θ, and a horizon H, a trajectory $T(x_0, \pi_\theta, H, I)$ is a sequence $(x_0, a_0, x_1, a_1, \ldots, x_{H-1},$ $a_{H-1}, x_H)$ such that $(x_0, \ldots, x_H) \in \mathcal{P}$ is an LS path (the multiset $\{x_0, \ldots, x_H\}$ will be denoted $P(T(x_0, \pi_\theta, H, I)))$, $\forall i \in [\![0, H-1]\!], a_i = \pi_\theta(o(x_i))$ and $\forall i \in [\![1, H]\!], x_i = \delta(x_{i-1}, a_{i-1})$. Note that the trajectory is defined with regard to solutions belonging to $\mathcal{X}$, while the policy operates on the space of observations obtained from these solutions. Compared to classic reinforcement schemes, where a reward can be assigned after each action, in our context, the reward will be computed globally for a given trajectory. Note that if we considered only the best-improvement strategy, then the reward could be assigned after each move to assess that the best move has been selected. Unfortunately, such a strategy will only be suitable for simple smooth unimodal problems. Hence we translate this reward as a function $R(x_0, \pi_\theta, H, I) = \max_{x \in P(T(x_0, \pi_\theta, H, I))} f_{\text{obj}}(x)$.

Example 1. In order to illustrate our framework, let us consider a basic hill climber (HC) that uses a simple best-improve strategy using objective function f_{obj} as fitness function. In the following, we only consider LS processes that do not involve memory that records past decisions. Hence, a state of a HC is fully described by a current solution $x \in \mathcal{X}$ and its neighborhood $\mathcal{N}(x)$. An observation will abstract the state as the variation of the fitness function for each possible flip of the values of x, $o(x) = (f_{\text{obj}}(flip_1(x)) - f_{\text{obj}}(x), \ldots, f_{\text{obj}}(flip_N(x))$ $f_{\text{obj}}(x)) \in \mathbb{R}^N$ ($\Omega = \mathbb{R}^N$). In a best-improve HC process, the search stops when a local optimum is reached and no improving move can be performed. We consider $\mathcal{A} = \{flip_i \,|\, i \in [\![1, N]\!]\} \cup \{Id\}$, where Id is the identity function on $\mathcal{X}$. Then, we define the policy $\pi_{HC}(o(x)) = \text{argmax}_{a \in \mathcal{A}}(f_{\text{obj}}(\delta(x, a)) - f_{\text{obj}}(x))$. Note that when a local optimum is reached, the identity action will always be selected for the remainder of the process.

Our objective is to maximize the maximum score encountered by the agent during its trajectory, and not the sum of local fitness improvements obtained during its trajectory. We are therefore not in the case of learning a Markov decision process. This is why classical reinforcement learning algorithms such as Q-learning or policy gradient are not applicable in this context.

This justifies our choice of neuro-evolution where the policy parameters will be abstracted by a neural network that will be used to select the suitable action. The parameters of this neural network will be searched by means of an evolutionary algorithm according to a learning process defined below.

2.3   Policy Learning for a Set of Instances

In this paper, we focus on NK landscapes as pseudo-boolean optimization problems. The NK landscape model was introduced to describe binary fitness landscapes [14]. The characteristics of these landscapes are determined by two key parameters: N, which represents the dimension (number of variables), and K (where $K < N$), which indicates the average number of dependencies per variable and, in turn, influences the ruggedness of the fitness landscape. An NK problem instance is an optimization problem represented by an NK function. We use random NK functions to create optimization problem instances with adjustable search landscape characteristics. This adjustment will be achieved by varying the parameters (N, K), thus generating diverse search landscapes. Hence, we consider $\mathcal{NK}(N, K)$ as a distribution of instances generated by a random NK function generator.

In order to achieve our policy learning process, we must assess the performance of a policy as $F(\pi_\theta, \mathcal{NK}(N, K), H) = \mathbb{E}_{I \sim \mathcal{NK}(N,K), x_0 \sim \mathcal{X}}[R(x_0, \pi_\theta, H, I)]$ where $x_0 \sim \mathcal{X}$ stands for a uniform selection of an element in $\mathcal{X}$. However, since this expectancy cannot be practically evaluated, we rely on an empirical estimator computed as an average of the score obtained by the policy π_θ for a finite number q of instances $I_1, \ldots, I_q$ sampled from the distribution $\mathcal{NK}(N, K)$ and a finite number r of restarts $x_0^{(1)}, \ldots, x_0^{(r)}$ drawn uniformly in $\mathcal{X}$:

$$\bar{F}(\pi_\theta, \mathcal{NK}(N, K), H) = \frac{1}{qr} \sum_{i=1}^{q} \sum_{j=1}^{r} R(x_0^{(j)}, \pi_\theta, H, I_i). \tag{1}$$

Figure 1 highlights our general learning methodology and the connections between the LS process at the instance solving level and the policy learning task that will be achieved by a neural network presented in the next section.

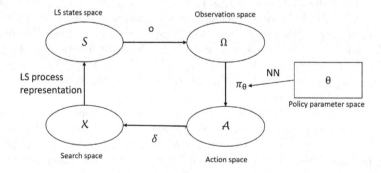

Fig. 1. Global View of the Process

3   Deterministic Local Search Policies for Pseudo-boolean Optimization

In this paper, our objective is to compare an LS policy learned by neuro-evolution, with three different baseline LS algorithms. To ensure a fair comparison among all the algorithms, all the different strategies take as input the same vector of observations corresponding to the variation of the objective/fitness function f for each possible flip of the values of x: $o(x) = (f(flip_1(x)) - f(x), \ldots, f(flip_N(x)) - f(x)) \in \mathbb{R}^N$. The set of possible actions available to the different strategies always remains $\mathcal{A} = \{flip_i \,|\, i \in [\![1, N]\!]\}$.

3.1   Neural Network Local Search Policy

We introduce a deterministic LS policy $\pi_\theta : \mathbb{R}^N \to \mathcal{A}$, called Neuro-LS, which uses a neural network g_θ, parametrized by a vector of real numbers θ. This neural network takes as input an observation vector $o \in \mathbb{R}^N$ and gives as output a vector $g_\theta(o)$ of size N whose component $g_\theta(o)_i$ corresponds to a preference score associated to each observation o_i. Then, the action a_i corresponding to the highest score $g_\theta(o)_i$ is selected.

Permutation Equivariant Neural Network

A desirable property of this neural network policy is to be *permutation equivariant* with respect to the input vector of observations, which is a property generally entailed by a local search algorithm, in order to make its behavior consistent for solving any type of instance. Formally, an LS algorithm is said *to be invariant to permutations* in the observations if for any permutation σ on $[\![1, N]\!]$, we have $a_{\sigma(i)} = \pi_\theta(o_{\sigma(1)}, o_{\sigma(2)}, \ldots, o_{\sigma(N)})$. As an example, the basic hill climber HC defined with Example 1 in Sect. 2.2 has this property.

In order to obtain this property for Neuro-LS, g_θ must be function from $\mathbb{R}^N$ to $\mathbb{R}^N$, such that for any permutation σ we have $(g_\theta(o)_{\sigma(1)}, \ldots, g_\theta(o)_{\sigma(N)}) = g_\theta(o_{\sigma(1)}, \ldots, o_{\sigma(N)})$. Such type of permutation equivariant neural network can be obtained by using the deep sets architecture [31].

Each layer of the proposed network combines the treatment of each observation o_i associated to each of the N variables with an additional operation that performs an average of the features across the different variables. This N-averaging operation is independent by permutation of the inputs and allows to *transmit* some general contextual information between the N features vector.

For a vector of observation $o = (o_1, o_2, \ldots, o_N)$ given as input, a permutation equivariant network g_θ with P hidden layers is defined as $g_\theta(o) = \phi_{\theta_P} \circ \phi_{\theta_{P-1}} \circ \cdots \circ \phi_{\theta_0}(o)$, where each ϕ_{θ_j} is a permutation invariant function from $\mathbb{R}^{N \times l_j}$ to $\mathbb{R}^{N \times l_{j+1}}$. The l_j values correspond to the layer sizes. Note that for the first layer $l_0 = 1$ and for the last layer $l_{P+1} = 1$. This network g_θ is shown in Fig. 2.

Each layer operation ϕ_{θ_j} with l_j input features and l_{j+1} output features includes a weight matrix $W_j \in \mathbb{R}^{l_j \times l_{j+1}}$ that treats each variable information independently, a variable-mixing weight matrix $\Gamma_j \in \mathbb{R}^{l_j \times l_{j+1}}$ and a bias vector

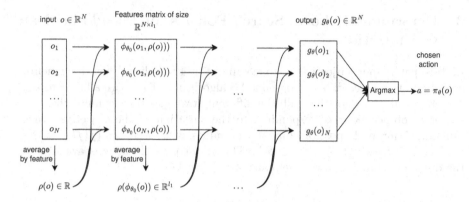

Fig. 2. Neural network policy $\pi_\theta : \Omega \to \mathcal{A}$ using a deep sets network architecture. Given a vector of observation $o \in \mathbb{R}^N$, the neural network outputs a vector $g_\theta(o) \in \mathbb{R}^N$. Then, the action $a = \operatorname{argmax}_{i \in [\![1,N]\!]} g_\theta(o)_i$ is returned by π_θ.

$\beta_j \in \mathbb{R}^l$. Note that the size of these matrices W_j and Γ_j does not depend on the size N of the observation vector, allowing the Neuro-LS strategy to adapt to pseudo-Boolean optimization problems of different sizes.

When taking as input a feature matrix $v = (v_1, \ldots, v_N)$ of size $N \times l_j$, the matrix of weights W_j processes each feature vector v_i associated with each variable i independently. Then, the weight matrix Γ_j processes an average vector $\rho(v)$ computed across the different N feature vectors of size l_j and given by $\rho(v) = \frac{1}{N} \sum_{i=1}^{N} v_i$.

Given $v \in \mathbb{R}^{N \times l_j}$, the output of the layer ϕ_{θ_j} is a matrix in $\mathbb{R}^{N \times l_{j+1}}$, which is the concatenation of N output vectors of size l_{j+1}, $\phi_{\theta_j}(v) = (\phi_{\theta_j}(v)_1, \ldots, \phi_{\theta_j}(v)_N)$, where for $1 \leq i \leq N$,

$$\phi_{\theta_j}(v)_i = \eta(\beta_j + x_i W_j + \rho(v) \Gamma_j), \tag{2}$$

with $\eta : \mathbb{R}^{l_j} \to \mathbb{R}^{l_{j+1}}$ the element-wise nonlinear activation map defined by $\eta(z) := (\tanh(z_1), \ldots, \tanh(z_{l_j}))$. We denote $\theta := \{(W_j, \Gamma_j, \beta_j)\}_{j \in [\![0,P]\!]}$, the set of all weight matrices and bias vectors of the neural network.

Neuro-Evolution with CMA-ES

The neural network policy π_θ is characterized by a set of parameters denoted as θ. The optimization goal is to maximize the estimated score $\bar{F}(\pi_\theta, \mathcal{NK}(N, K), H)$. This poses a stochastic black-box optimization problem within the real-valued search space $\mathbb{R}^{|\theta|}$. To tackle this problem, we propose to use the covariance matrix adaptation evolution strategy (CMA-ES) [8], which stands out as one of the most powerful evolutionary algorithms for addressing such black box optimization problems [18].

The principle of CMA-ES is to iteratively test new generations of real-valued parameter vectors θ (individuals). Each new generation of parameter vectors is stochastically sampled according to a multivariate normal distribution. The mean and covariance matrix of this distribution are incrementally updated, so as to maximize the likelihood of previously successful candidate solutions. Thanks to the use of a stepwise-adapted covariance matrix, the algorithm is able to quickly detect correlations between parameters, which is an important advantage when optimizing the (many) parameters of the neural network policy. Another advantage of CMA-ES is that it relies on a ranking mechanism of the estimated scores $\bar{F}$ given by the Eq. (1) for the different individuals of the population, rather than on their absolute values, making the algorithm more robust to stochastic noise related to the incertitude on the estimation of the performance score F with a finite number of trajectories.

3.2   Basic Local Search Policies

We compare the neural network policy above with three *basic* local search policies which have been extensively studied in the literature [1,19]. All these strategies take as input the same vector of observation as the neural network policy and return an action $a \in \mathcal{A}$. These three policies are two hill climbers as well as a $(1,\lambda)$-evolution strategy [3] used as a local search [25]. They are made deterministic using a pseudo-random number generator h whose seed is determined with a hash function from the current state x encountered by the LS.

Best Improvement Hill Climber [+jump] (BHC$^+$). This strategy always selects the action $a_i = flip_i \in \mathcal{A}$ in such a way that $f(a_i(x)) - f(x)$ is maximized, provided there is at least one action a_i that strictly improves the score. If $\forall i, \ f(a_i(x)) - f(x) \leq 0$, then this strategy performs a random jump by choosing a random action $a \in \mathcal{A}$ using the pseudo-random number generator $h(x)$.

First Improvement Hill Climber [+jump] (FHC$^+$). This strategy iterates through all actions in $\mathcal{A}$ in random order and selects the first action a_i leading to a strictly positive score improvement, i.e. such that $f(a_i(x)) - f(x) > 0$. Similar to the BHC$^+$ strategy, if $\forall i, \ f(a_i(x)) - f(x) \leq 0$, it performs a random jump.

(1,λ)-Evolution Strategy ((1,λ)-ES). This strategy randomly evaluates λ actions in $\mathcal{A}$ and chooses the one that yields the best score, even if it results in a deteriorating move. λ is a method hyperparameter that will be calibrated to maximize the estimated score $\bar{F}$ for each type of NK landscape instance as detailed in the next section.

4   Computational Experiments

The aim of this section is to answer two questions experimentally. The first concerns the performance of Neuro-LS compared to the baseline LS strategies presented in the last subsection for NK landscape problems of different size and ruggedness. The second is to study the emergent strategies discovered by Neuro-LS at the end of its evolutionary process.

First, we discuss the experimental setting. Then, we follow the classical steps of a machine learning workflow: Subsect. 4.2 describes the Neuro-LS training process; it will allow us to select different emerging strategies for each type of NK landscape on validation sets, which we will then compare with the different baseline LS strategies on test sets (Subsect. 4.3). Finally, an in-depth analysis of the best emerging strategies discovered by Neuro-LS will be performed in Subsect. 4.4.

4.1   Experimental Settings

In these experiments, we consider independent instances of NK-landscape problems. 12 different scenarios with $N \in \{32, 64, 128\}$ and $K \in \{1, 2, 4, 8\}$ are considered. For each scenario, three different sets of instances are sampled independently from the $\mathcal{NK}(N, K)$ distribution described in Sect. 2.3: a training set, a validation set and a test set. For the resolution of each instance, given a random starting point $x_0 \in \mathcal{X}$, each LS algorithm performs a trajectory of size $H = 2 \times N$ (iterations) and returns the best solution found during this trajectory. The experiments were performed on a computer equipped with a 12th generation Intel®CoreTM i7-1265U processor and 14.8 GB of RAM.

Neuro-LS is implemented in Python 3.7 with Pytorch 1.4 library.[1] For all experiments with different values N and K, we use the same architecture of the neural network composed of two hidden layers of size 10 and 5, with a total of $|\theta| = 162$ parameters to calibrate. To optimise the weights of the neural network, we used the CMA-ES algorithm of the pycma library [7]. The multivariate normal distribution of CMA-ES is initialized with mean parameter μ (randomly sampled according to a unit normal distribution) and initial standard deviation $\sigma_{init} = 0.2$.

4.2   Neuro-LS Training Phase

For each NK-landscape configuration, we run 10 different training processes of Neuro-LS with CMA-ES, with a time limit of two hours, to optimize the empirical score $\bar{F}$ defined by Eq. (1) computed as an average of the best fitness scores obtained for 50 trajectories ($r = 5$ independent random restarts for each of the $q = 10$ training instances).

Figure 3 displays the results of 10 independent neuro-evolution training processes for the NK landscape instances with $N = 32$ and $K = 8$. At each generation, the 10 learning instances are regenerated to avoid over-fitting, then CMA-ES samples a population of 19 individuals (19 vectors of weights $\theta \in \mathbb{R}^{162}$ of the neural network), and the best Neuro-LS strategy of the population on the training set is evaluated on the 10 instances of the validation set.

The evolution of the average score of Neuro-LS on the training and validation sets are indicated with orange and blue lines in Fig. 3. The red line is a reference

---

[1] The program source code, benchmark instances and result files are available at the url https://github.com/Salim-AMRI/NK_Landscape_Project.git.

score. It corresponds to the average score $\bar{F}$ obtained by the BHC$^+$ local search strategy (see Sect. 3.2) on the same validation set.

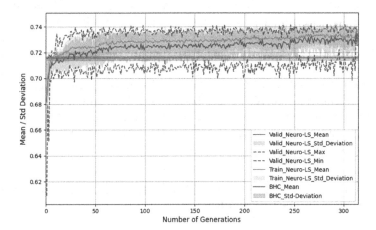

Fig. 3. Evolution of the average score obtained by Neuro-LS on the training and validation sets over the generations of CMA-ES.

First, we find that the validation curve closely follows the training curve in average, even if it is slightly below. This reassuring consistency suggests that our model successfully generalizes the strategy learned in training instances.

Furthermore, upon comparing the validation curves of Neuro-LS and BHC$^+$, we observe that the Neuro-LS curve progresses over generations, and eventually surpasses BHC$^+$ when evaluated on the same set of validation instances. This finding highlights that, once trained, the Neuro-LS method is able to find solutions more efficiently compared to the baseline BHC$^+$ local search for these types of instances.

In Fig. 3, the minimum and maximum scores on the validation set obtained during the 10 independent neuro-evolution runs are also indicated with blue dashed lines. The figure illustrates significant variability in the results, highlighting the diversity in the performance of the emerging strategies. Nevertheless, this variability poses no issue in our context, as only the strategy with the best results on the validation set will be selected for the test phase presented in the next subsection.

4.3   Test Phase

In this phase, we performed a series of evaluations to assess whether the best Neuro-LS strategies, selected based on the validation set for each configuration of NK landscape, continue to perform well in new test instances that are independently sampled from the same $\mathcal{NK}(N, K)$ distribution.

Table 1 summarizes the average score obtained by Neuro-LS and the three other competing methods, namely the BHC$^+$, FHC$^+$ and (1, λ)-ES algorithms, presented in Sect. 3.2. The strategy (1, λ)-ES has one hyperparameter λ that we calibrated in the range $[\![1, N]\!]$ on training instances for each (N, K) configuration.

In order to perform a fair comparison between the different strategies, we computed an average estimated score $\bar{F}$ on the same 100 test instances. For each instance we use the same starting point to compute the trajectory produced by each LS strategy.[2]

In Table 1, the best average result obtained among the mentioned methods for each configuration (N, K) of test instances is highlighted in bold. Results underlined indicate significant better results in average compared to all the other strategies (p-value below 0.001), measured with a Student t-test without assuming equal variance.[3]

Table 1. Average results on test instances obtained by different local search strategies for different NK landscape configurations.

Instances		Methods			
N	K	BHC$^+$	FHC$^+$	(1, λ)-ES	Neuro-LS
32	1	0.694	0.688	0.695	**0.699**
32	2	0.713	0.709	0.717	**0.721**
32	4	0.717	0.721	0.713	**<u>0.735</u>**
32	8	0.702	0.712	0.707	**<u>0.732</u>**
64	1	0.700	0.693	0.696	**0.702**
64	2	0.712	0.709	0.712	**0.715**
64	4	0.721	0.721	0.714	**<u>0.734</u>**
64	8	0.710	0.719	0.707	**<u>0.737</u>**
128	1	0.698	0.691	0.696	**0.701**
128	2	0.712	0.711	0.710	**0.713**
128	4	0.724	0.723	0.717	**0.728**
128	8	0.711	0.719	0.705	**<u>0.730</u>**

Table 1 shows that Neuro-LS always obtains better results for all the configurations of NK landscape, but the difference in score is only really significant when $K = 4$ and $K = 8$ for all values of N (except for $N = 128$ and $K = 4$). It

---

[2] For this evaluation test, we only perform one restart per instance, to avoid any dependency between the different executions that might take place on the same instance. It allows to obtain a distribution of 100 independently and identically distributed scores for each strategy and for each NK configuration.

[3] The normality condition required for this test was first confirmed using a Shapiro statistical test on the empirical distributions of 100 iid scores obtained by each strategy.

means that our strategy becomes more effective than other methods when the landscape is more rugged.

This score improvement compared to the other baseline methods, obtained with the same budget of $H = 2 \times N$ iterations performed on each instance, can be attributed to a more efficient exploration of the search space as K increases (as seen in the next subsection).

4.4   Study of Neuro-LS Emerging Strategies

The objective here is to analyse in detail the best Neuro-LS strategies that have emerged with neuro-evolution and to understand their decision-making processes for the different types of landscape studied (smooth or rugged).
We have observed two main patterns of emerging strategies:

- For smooth landscapes, when $K = 1$ or $K = 2$, Neuro-LS learns to perform almost all the time a best improvement move, which explains why for these instances, it obtains almost the same score as the BHC$^+$ strategy (see Table 1).
- For rugged landscapes, when $K = 4$ and $K = 8$, the emerging strategy is much more interesting. Figure 4 displays a representative example of the trajectory performed by Neuro-LS when $N = 64$ and $K = 8$. This figure shows two graphs based on data collected during the resolution of this instance. The graph on the top of this Figure shows the evolution of the fitness reached by Neuro-LS over the $H = 2 \times N = 128$ iterations. The graph below shows at each iteration the number of available actions corresponding to an improvement of the score (in blue), and the rank of the action selected by Neuro-LS, measured in term of fitness improvement (in red). A rank of 1 on this plot indicates that Neuro-LS has chosen a best improvement move, while a rank of 64 indicates a worst deteriorating move (note that the y-axis is inverted, because it is a maximization problem). We observe on this plot that the emerging Neuro-LS strategy has three different successive operating modes:

  1. **Median hill climbing behavior.** When the number of actions associated with a positive improvement of the score, N_a^+, is relatively large ($\sim$ above $N/10$), Neuro-LS does not always choose the best improvement move, but instead a move with a rank approximately equal to $N_a^+/2$. This provides a good compromise between improving the score and avoiding being trapped too quickly in a local optimum.
  2. **Best improvement hill climbing behavior.** When the number of actions corresponding to a positive delta fitness is low ($\sim$ below $N/10$), Neuro-LS often chooses the best move (with rank 1) to quickly converge toward the closest local optimum.
  3. **Jump with worst move.** When there is no more improving move, Neuro-LS does not stagnate, but instead directly chooses to perform the worst possible move (with rank 64). Even if this movement considerably deteriorates the current fitness score, it actually maximizes its long-term chances of escaping the current local optimum and continually exploring

new areas of the search space. Indeed, we observe on this plot that Neuro-LS continuously improves its score with this strategy for this instance. Note that after choosing the worst possible move, it does not choose the best possible move, otherwise it would return to the same local optimum.

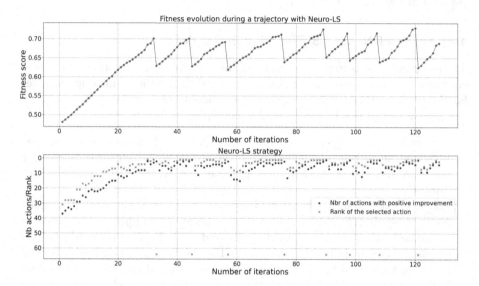

Fig. 4. Fitness evolution curve and strategy used by Neuro-LS for the resolution of a instance with significantly rugged NK landscape ($N = 64$ and $K = 8$).

5 Conclusion

Our study explores the emergence of new local search algorithms with neuro-evolution. Results on NK landscapes show that different neural network policies are learned, each adapted to the resolution of a particular landscape distribution type (smooth or rugged). Our algorithm is competitive with basic deterministic local search procedures for all the NK landscape types considered in this work. Particularly for rugged landscapes, it can achieve significantly better results with an original emerging strategy, using a worst-case improvement move, which proves very effective in the long run for escaping local optima.

This study outlines avenues for future research on the automatic discovery of more advanced strategies using as input a richer set of observations to make its decision. The proposed framework could also be applied to study the emergence of strategies adapted to other types of combinatorial optimization problems.

Acknowledgment. This work was granted access to the HPC resources of IDRIS (Grant No. AD010611887R1) from GENCI. The authors would like to thank the Pays de la Loire region for its financiel support for the Deep Meta project (Etoiles Montantes en Pays de la Loire). The authors also acknowledge ANR - FRANCE (French National Research Agency) for its financial support of the COMBO project (PRC - AAPG 2023 - Axe E.2 - CE23). We are grateful to the reviewers for their comments.

References

1. Basseur, M., Goëffon, A.: Hill-climbing strategies on various landscapes: an empirical comparison. In: Proceedings of the 15th Annual Conference on Genetic and Evolutionary Computation, pp. 479–486 (2013)

2. Basseur, M., Goëffon, A.: Climbing combinatorial fitness landscapes. Appl. Soft Comput. **30**, 688–704 (2015)

3. Beyer, H.G.: The Theory of Evolution Strategies. Springer, Berlin, Heidelberg (2001). https://doi.org/10.1007/978-3-662-04378-3

4. Cappart, Q., Chételat, D., Khalil, E.B., Lodi, A., Morris, C., Velickovic, P.: Combinatorial optimization and reasoning with graph neural networks. J. Mach. Learn. Res. **24**, 130:1–130:61 (2023)

5. Falkner, J.K., Thyssens, D., Bdeir, A., Schmidt-Thieme, L.: Learning to control local search for combinatorial optimization. In: Amini, MR., Canu, S., Fischer, A., Guns, T., Kralj Novak, P., Tsoumakas, G. (eds.) Machine Learning and Knowledge Discovery in Databases. ECML PKDD 2022. LNCS, vol. 13717, pp. 361–376. Springer, Cham (2023). https://doi.org/10.1007/978-3-031-26419-1_22

6. Glover, F.: Tabu search-part i. ORSA J. Comput. **1**(3), 190–206 (1989)

7. Hansen, N., Akimoto, Y., Baudis, P.: CMA-ES/pycma on Github. Zenodo (2019). https://doi.org/10.5281/zenodo.2559634

8. Hansen, N., Ostermeier, A.: Completely derandomized self-adaptation in evolution strategies. Evol. Comput. **9**(2), 159–195 (2001)

9. Hoos, H.H.: Programming by optimization. Commun. ACM **55**(2), 70–80 (2012)

10. Hoos, H.H., Stützle, T.: Stochastic Local Search: Foundations and Applications. Elsevier, Amsterdam (2004)

11. Hopfield, J.: Neural networks and physical systems with emergent collective computational abilities. Proc. Natl. Acad. Sci. U.S.A. **79**, 2554–8 (1982)

12. Hudson, B., Li, Q., Malencia, M., Prorok, A.: Graph neural network guided local search for the traveling salesperson problem. In: The Tenth International Conference on Learning Representations, ICLR 2022, Virtual Event, 25–29 April 2022. OpenReview.net (2022)

13. Jones, T., Forrest, S., et al.: Fitness distance correlation as a measure of problem difficulty for genetic algorithms. In: ICGA, vol. 95, pp. 184–192 (1995)

14. Kauffman, S.A., Weinberger, E.D.: The NK model of rugged fitness landscapes and its application to maturation of the immune response. J. Theor. Biol. **141**(2), 211–245 (1989)

15. Lourenço, H.R., Martin, O.C., Stützle, T.: Iterated local search. In: Glover, F., Kochenberger, G.A. (eds.) Handbook of Metaheuristics. International Series in Operations Research and Management Science, LNCS, vol. 57, pp. 320–353. Springer, Boston, MA (2003). https://doi.org/10.1007/0-306-48056-5_11

16. Malan, K.M.: A survey of advances in landscape analysis for optimisation. Algorithms **14**(2), 40 (2021)

17. Mamaghan, M.K., Mohammadi, M., Meyer, P., Karimi-Mamaghan, A.M., Talbi, E.: Machine learning at the service of meta-heuristics for solving combinatorial optimization problems: a state-of-the-art. Eur. J. Oper. Res. **296**(2), 393–422 (2022)
18. Müller, N., Glasmachers, T.: Challenges in high-dimensional reinforcement learning with evolution strategies. In: Auger, A., Fonseca, C.M., Lourenço, N., Machado, P., Paquete, L., Whitley, D. (eds.) PPSN 2018. LNCS, vol. 11102, pp. 411–423. Springer, Cham (2018). https://doi.org/10.1007/978-3-319-99259-4_33
19. Ochoa, G., Verel, S., Tomassini, M.: First-improvement vs. best-improvement local optima networks of NK landscapes. In: Schaefer, R., Cotta, C., Kołodziej, J., Rudolph, G. (eds.) PPSN 2010. LNCS, vol. 6238, pp. 104–113. Springer, Heidelberg (2010). https://doi.org/10.1007/978-3-642-15844-5_11
20. Özcan, E., Bilgin, B., Korkmaz, E.E.: A comprehensive analysis of hyper-heuristics. Intell. Data Anal. **12**(1), 3–23 (2008)
21. Santana, Í., Lodi, A., Vidal, T.: Neural networks for local search and crossover in vehicle routing: a possible overkill? In: Cire, A.A. (eds.) Integration of Constraint Programming, Artificial Intelligence, and Operations Research. CPAIOR 2023. LNCS, vol. 13884, pp. 184–199. Springer, Cham (2023). https://doi.org/10.1007/978-3-031-33271-5_13
22. Schiavinotto, T., Stützle, T.: A review of metrics on permutations for search landscape analysis. Comput. Oper. Res. **34**(10), 3143–3153 (2007)
23. Sörensen, K., Glover, F.: Metaheuristics. Encycl. Oper. Res. Manag. Sci. **62**, 960–970 (2013)
24. Talbi, E.: Machine learning into metaheuristics: a survey and taxonomy. ACM Comput. Surv. **54**(6), 129:1–129:32 (2022)
25. Tari, S., Basseur, M., Goëffon, A.: On the use of $(1, \lambda)$-evolution strategy as efficient local search mechanism for discrete optimization: a behavioral analysis. Nat. Comput. **20**, 345–361 (2021)
26. Trafalis, T.B., Kasap, S.: Neural networks for combinatorial optimization. In: Floudas, C., Pardalos, P. (eds.) Encyclopedia of Optimization, Second Edition, pp. 2547–2555. Springer, Boston, MA (2008). https://doi.org/10.1007/978-0-387-74759-0_439
27. Veerapen, N., Hamadi, Y., Saubion, F.: Using local search with adaptive operator selection to solve the progressive party problem. In: Proceedings of the IEEE Congress on Evolutionary Computation, CEC 2013, Cancun, Mexico, 20–23 June 2013, pp. 554–561. IEEE (2013)
28. Vuculescu, O., Pedersen, M.K., Sherson, J.F., Bergenholtz, C.: Human search in a fitness landscape: how to assess the difficulty of a search problem. Complexity **2020** (2020)
29. Whitley, D.: MK landscapes, NK landscapes, MAX-kSAT: a proof that the only challenging problems are deceptive. In: Proceedings of the 2015 Annual Conference on Genetic and Evolutionary Computation, pp. 927–934 (2015)
30. Willmes, L., Bäck, T., Jin, Y., Sendhoff, B.: Comparing neural networks and kriging for fitness approximation in evolutionary optimization. In: Proceedings of the IEEE Congress on Evolutionary Computation, CEC 2003, Canberra, Australia, 8–12 December 2003, pp. 663–670. IEEE (2003)
31. Zaheer, M., Kottur, S., Ravanbakhsh, S., Poczos, B., Salakhutdinov, R.R., Smola, A.J.: Deep sets. Adv. Neural Inf. Process. Syst. **30** (2017)

Q-Learning Based Framework for Solving the Stochastic E-waste Collection Problem

Dang Viet Anh Nguyen[1], Aldy Gunawan[1]([✉]), Mustafa Misir[2],
and Pieter Vansteenwegen[3]

[1] School of Computing and Information Systems, Singapore Management University,
Singapore, Singapore
{dvanguyen,aldygunawan}@smu.edu.sg

[2] Division of Natural and Applied Sciences, Duke Kunshan University, Kunshan,
China
mustafa.misir@dukekunshan.edu.cn

[3] KU Leuven Institute for Mobility - CIB, University of Leuven, Celestijnenlaan 300,
3001 Leuven, Belgium
pieter.vansteenwegen@kuleuven.be

Abstract. Electrical and Electronic Equipment (EEE) has evolved into a gateway for accessing technological innovations. However, EEE imposes substantial pressure on the environment due to the shortened life cycles. E-waste encompasses discarded EEE and its components which are no longer in use. This study focuses on the e-waste collection problem and models it as a Vehicle Routing Problem with a heterogeneous fleet and a multi-period planning problem with time windows as well as stochastic travel times. Two different Q-learning-based methods are designed to enhance the search procedure for finding solutions. The first method involves utilizing the state-action value to determine the order of multiple improvement operators within the GRASP framework. The second one involves a hyperheuristic that extracts a stochastic policy to select heuristic operators during the search. Computational experiments demonstrate that both methods perform competitively with state-of-the-art methods in newly-generated small-sized instances, while the performance gap widens as the size of the problem instances increases.

1 Introduction

Climate change is a global challenge, and all United Nations (UN) Members made commitments under the UN's 2030 Sustainable Development Agenda and Paris Agreement. Electronic goods, such as computers, laptops, and mobile phones, have a carbon footprint and directly contribute to global warming. Once they are no longer in use or nearing the end of their useful life, electronic waste (or e-waste) is generated [1]. Recently, there is an increased interest in the planning of e-waste collection [2]. The process needs to be planned carefully, otherwise it may contribute to congestion and worsen urban air pollution [3]. Pourhejazy

© The Author(s), under exclusive license to Springer Nature Switzerland AG 2024
T. Stützle and M. Wagner (Eds.): EvoCOP 2024, LNCS 14632, pp. 49–64, 2024.
https://doi.org/10.1007/978-3-031-57712-3_4

et al. [4] describes a study of planning vehicle routes for the e-waste collection problem and formulates the problem as a variant of the Vehicle Routing Problem (VRP).

E-waste can be collected either from fixed sites, namely e-bin, or from the doorstep of households. Gunawan et al. [5] formulated the e-waste collection problem as a deterministic VRP with a heterogeneous fleet and a multi-period planning problem with time windows. A hybrid metaheuristic, based on a Greedy Randomized Adaptive Search Procedure reinforced by Path-Relinking (GRASP-PR), was proposed to solve newly generated instances [6]. [4] introduced an original capacitated general routing with time-window model by considering an integrated collection scheme that simultaneously accommodates on-call and door-to-door demands. [7] presented a VRP model with intermediate facilities and heteregeneous vehicles for e-waste collection encompassing both households and mobile collection. In this work, the model from [5] is extended by incorporating stochastic travel times. Therefore, we call the problem the Heterogeneous VRP with Multiple Time Windows and Stochastic Travel Time (HVRP-MTWSTT).

Machine Learning techniques have recently gained more attentions for solving Combinatorial Optimization Problems [8,9]. One common technique is Reinforcement Learning (RL) [10]. RL agents are designed for dynamic environments. The RL component primarily serves as a Credit Assignment (CA) mechanism for adaptive operator selection within a metaheuristic framework [11,12]. Several studies have employed RL techniques to develop adaptive search operator selection within the search procedure for solving VRP [13,14]. The Q-learning algorithm [15] falls under the category of temporal-difference (TD) learning in the realm of RL. The state-value or state-action value (denoted as Q, representing the expected cumulative reward an agent can obtain by taking a specific action from a given state in a Markov Decision Process (MDP)) is updated based on other learned estimates through bootstrapping.

In this study, two Q-learning-based methods are designed to enhance the search procedure for finding solutions. The former involves using the state-action value (Q-value) to determine the order of multiple improvement operators within the GRASP framework, referred to as GRASP-Q. The latter employs a hyperheuristic that extracts a stochastic policy from the Q-table (a tabular structure used to store Q-values) to select heuristic operators during the search, known as HH-Q. In our comparative analysis with GRASP-PR [6] and SimGRASP [16], we observed that a hyperheuristic with an adaptive operator selection framework, employing a stochastic policy based on a trained Q-table, achieves superior performance. Using the trained Q-table to fine-tune the order of operators within the GRASP metaheuristic could enhance its performance. The performance gap between HH-Q and the other baselines increases as the problem size grows, making the method with an adaptive operator selection mechanism based on learning methods particularly advantageous for practical-sized instances.

The paper outline is as follows. Section 2 describes the problem. Section 3 explains the proposed algorithms. Section 4 reports the experimental setup and computational results. Finally, Sect. 5 presents the conclusions, limitations of the study and suggestions for possible future research directions.

2   Problem Description

The problem in this paper, HVRP-MTWSTT, considers two types of collection: fixed drop-off points (e-bins) and customers' on-demand requests. The former are fixed points that need to be visited only when they reach a specified level, while the latter involve collections at households and are associated with time windows. The collection task typically spans several days, with specific destinations requiring collections. Each customer has preferred days and time window, resulting in multiple time windows. E-bins are available throughout the collection period until they are reached by a collector. For each on-demand request, a fee is charged to the customer. However, due to the impact of arrival times on customer satisfaction, if a vehicle arrives after the upper bound of the time windows on that day, a penalty cost is imposed. Conversely, arriving earlier than the on-demand request time windows results in an idle fee. A heterogeneous fleet with two types of vehicles is employed for the collection task. Each type of vehicle has a different capacity and operating cost per unit of time. Service times at each collection point vary due to differences in e-waste demand.

The deterministic formulation (HVRP-MTW) has been presented in [5]. Here, we extend it by considering stochastic travel times that occur in an urban context where various factors, such as traffic congestion and unexpected events, can influence travel times. The objective of the problem is to maximize the profit of the e-waste collection process, considering factors such as revenue from customer on-demand collection requests, vehicle operating costs, potential idle and penalty costs, while simultaneously adhering to constraints related to vehicle capacity, vehicle limitations, and collection point period constraints. The travel time between the depot and collection points, as well as between two collection points, follows a uniform distribution as follows:

$$\tilde{t}_{ij} = d_{ij} \frac{\theta}{100} \tag{1}$$

where $\tilde{t}_{ij}$ represents the stochastic travel time between the depot and collection points and between two collection points. d_{ij} denotes the Euclidean distance between any two points i and j on the graph. $\frac{\theta}{100}$ is a normalized noise term with θ following a discrete distribution $\mathcal{D}\{1, 100\}$. The travel time $\tilde{t}_{ij}$ is only discovered after the vehicle traverses from point i to point j.

3   Integrating Q-Learning to Enhance Search Procedures

3.1   Modeling Search Process as a MDP

We introduce two different Q-learning-based methods, namely GRASP-Q and HH-Q. Both methods require a trained Q-table as input. The search process for the HVRP-MTWSTT is modelled as an MDP, where an agent can interact with the environment to receive feedback on which action to take in order to maximize the return. Specifically, the agent will learn to decide which operator to select in the next step to continuously improve the current solution. The MDP is a sequential decision-making framework that typically includes various components, such as states S, actions A, rewards R, policy $(\pi(\theta))$, and transition functions. The following sections discuss these MDP components in detail.

State Space. The state refers to the current state of the environment in which the agent makes decisions. It contains relevant information that the agent needs to make decisions. Tabular-based learning methods like Q-learning are ineffective in handling large state spaces. Here, we only retain a simple state, i.e., the action taken in the previous step, denoted as a_{t-1}. As a result, by considering the state as an action, which action to take after the current action can be determined. This forms the basis of our approach for establishing the order of actions within GRASP and selecting the next action within HH-Q.

Actions. In the proposed MDP framework, the actions represent problem-dependent heuristics that are components of the GRASP metaheuristic framework. These heuristics, used as actions, can be either local search operators aimed at improving the current solution or generation heuristics designed to create a new solution to escape local optima. We employ the components of GRASP-PR, which was introduced to solve the deterministic HVRP-MTWSTT [6], as actions within the MDP framework. Action 0 in the MDP framework corresponds to a constructive heuristic responsible for generating an initial solution for the problem. Here, it creates a solution by utilizing the Clarke and Wright (CW) heuristic with Realistic Opportunity Savings - γ (ROS-γ) approach for a heterogeneous vehicle fleet [17]. It incorporates the fixed costs of different vehicles in the fleet when calculating the savings from combining two subtours. Additionally, a repair operator has been designed to ensure that the created solution adheres to the collection period constraint [6].

Actions 1 to 11 represent various local search operators. These include: relocating two nodes, swapping two nodes, employing 2-opt, or-opt, 3-opt, adding a new route, creating a new route from two randomly selected nodes from two existing routes, and combining two existing routes. Detailed descriptions of these operators can be found in [6]. Action 12 involves the path-relinking procedure (Algorithm 1). The path-relinking process takes as input the current initial solution x_i and a guide solution x_g selected from a solution pool that stores elite solutions obtained during the search procedure (Line 1). The algorithm then initializes the best solution in the path between these solutions in Lines 2–3. The symmetric difference Δ between the initial solution and the guide solution is calculated, storing the indices of the nodes where x_i differs from x_g (Line 5). The relinking process is executed by swapping nodes in the initial solution x_i to transform it into the guiding solution x_g (Lines 6–9). Any solution found during the relinking process, denoted as x', is validated to ensure it is a valid solution before being compared with the best solution in the relinking process (Line 10). If the newly found solution is better than the best solution found so far, it becomes the best solution in the next iteration (Lines 11–13). The relinking process terminates when the current solution x reaches the guiding solution x_g, and it returns the best solution found in the relinking path (Line 17).

Algorithm 1. Path-relinking

1: **Input:** x_i, x_g
2: $x \leftarrow x_i$;
3: $f(x^*) \leftarrow \max\{f(x_i), f(x_g)\}$;
4: $x^* \leftarrow \arg\max\{f(x_i), f(x_g)\}$;
5: Symmetric difference, $\Delta(x, x_g) \leftarrow i : x[i] \neq x_g[i]$;
6: **while** $\Delta(x, x_g) \neq \emptyset$ **do**
7:     An index $t \in \Delta(x, x_g)$ where $x[t] \neq x_g[t]$;
8:     Selects a node $i = x[t]$ and $j = x_g[t]$, $i, j \in x$;
9:     $x' \leftarrow$ Swap two selected nodes, $\text{swap}(x_i, i, j)$;
10:    **if** $\text{valid}(x') = \text{True}$ **then**
11:        **if** $f(x') > f(x^*)$ **then**
12:            $x^* \leftarrow x'$;
13:            $f(x^*) \leftarrow f(x')$;
14:        **end if**
15:    **end if**
16: **end while**
17: **Return:** $x^*, f(x^*)$;

Reward Function. The reward function plays a crucial role in providing feedback to the agent after it takes an action. Previous research has made use of the ALNS scoring system [18] or normalized the difference between the current solution and the newly found solution [14]. However, in our MDP framework, the action set contains action 0, which corresponds to a constructive heuristic typically producing a low-quality solution. When applying any action from 1 to 12 to the solution generated by action 0, it is expected to yield a better solution. Consequently, if we incentivise the agent to prefer the newly discovered solution over the current one, following the ALNS scoring mechanism, the agent might exploit this reward system by consistently selecting a combination of action 0 and another improvement operator instead of diversifying its search to find better solutions. Therefore, we have modified the reward function by simplifying the ALNS scoring system to include only one rewarding criterion:

$$r(s_t, s_{t+1}) = \begin{cases} 5, & \text{if } f(x') > f(x^*) \\ 0, & \text{Otherwise.} \end{cases} \tag{2}$$

3.2 Q-Learning Algorithm

Q-learning is an off-policy Temporal-Difference control method in which each state-action value of state S_t and action A_t at time step t ($Q(S_t, A_t)$) are indicated in a tabular. In each step, the agent selects its action based on a specific ϵ-greedy policy. Under this policy, the agent typically behaves greedily by choosing the action with the maximum $Q(S_t, A_t)$ value, given the current state s_t. However, with a small probability ϵ, the agent randomly selects an action from the set of all available actions, each having an equal probability of being chosen. The agent then updates the Q value of the current state S_t based on the maximum of the Q-values of the next state-action pair. The update rule of Q-learning is:

$$Q(S_t, A_t) \leftarrow Q(S_t, A_t) + \alpha \left[R_{t+1} + \gamma \max_a Q(S_{t+1}, a) - Q(S_t, A_t) \right] \tag{3}$$

where $Q(S_t, A_t)$ is the state-action value when taking action A_t at state S_t. $\alpha \in (0, 1]$ is the learning rate, whereas γ is the discount factor. R_{t+1} is the reward received after taking action A_t at state S_t.

The Q-learning algorithm is presented in Algorithm 2. First, all the algorithm parameters, namely α, γ, ϵ, and the number of training episodes n, need to be set. The state-action value table is then initialized (Lines 1–5). For each training episode, the initial state S is established (Line 7). During each step of this episode, the algorithm selects an action A based on the current state S, following an ϵ-greedy policy. The algorithm then takes action A, receives the reward R, and observes the next state S' (Line 10). Subsequently, the action-state value corresponding to state S and action A, $Q(S, A)$ is updated based on the reward R and the next state S' using Eq. 3 (Line 11). The next state S' then becomes the current state for the next step, and this process is iterated until a terminal state is reached (Lines 12–13).

Algorithm 2. Q-learning algorithm

1: **Input:** Learning rate $\alpha \in (0, 1]$;
2: Discounted factor, γ
3: Exploration rate, $\epsilon > 0$;
4: Training episodes, n;
5: Initialize $Q(s, a)$ for all $s \in \mathcal{S}$, $a \in \mathcal{A}$;
6: **for** $i = 0$ to n **do**
7:    Initialize state, S;
8:    **for** Each step in episode i **do**
9:       Based on state S, choose action A using $\epsilon - greedy$ policy;
10:       Take action A, observe R, S';
11:       $Q(S, A) \leftarrow Q(S, A) + \alpha \left[R + \gamma \max_a Q(S', a) - Q(S, A) \right]$;
12:       $S \leftarrow S'$;
13:       Until S is terminal
14:    **end for**
15: **end for**

3.3   Training Algorithm

First, a Q-table to store the state-action values is created (Line 3 - Algorithm 3). In each training episode, the algorithm initializes a problem instance and a feasible solution x for the problem using action 0. It also initializes the best solution, the initial state, and the solution pool for the path-relinking process (Lines 6–10). Next, based on the initial state and an ϵ-greedy policy, the agent selects an action a. This action is a heuristic operator that applies to the current solution in order to explore a new solution x', as well as to obtain a reward r and a new state s' (Lines 12–15).

If the new solution x' satisfies the condition to enter the solution pool P, it will be added to the pool P (Lines 16–17). If the taken action is action 0, the algorithm will restart the search process (Lines 19–20). The newly found solution

x' is then compared with the best solution so far, denoted as x^*. If x' is better than x^*, it becomes the new best solution x^* in the next step (Lines 22–23). The state-action value corresponding to action a and state s, $Q(s, a)$ is updated by Eq. 3, and the new state s' is set to the state s for the next step (Lines 25–26). At the end of the training process, the algorithm returns the Q-table, which stores the state-action values for all state-action pairs (Line 29).

Algorithm 3. Training algorithm

1: **Input:** Number of training episodes, M
2: Number of steps, m
3: Initialize Q-table
4: $i = 0$
5: **while** $i < M$ **do**
6:     Initialize a problem instance;
7:     Create a feasible solution, x;
8:     Initialize best solution, x^*, $f(x^*) \leftarrow x, f(x)$
9:     Initialize initial state, $s = 0$
10:     Initialize pool P for Path-relinking
11:     **for** step $= 0$ to m **do**
12:         $a \leftarrow$ action given by $\epsilon - greedy$ policy for state s;
13:         Take action a
14:         Observe new solution x'
15:         Observe reward r and new state s'
16:         **if** $f(x') > (1 - \mu)f(x^*)$ **then**
17:             Add x' to P
18:         **end if**
19:         **if** action a creates new solution **then**
20:             $x, f(x) = x', f(x')$
21:         **end if**
22:         **if** $f(x') > f(x^*)$ **then**
23:             $x^*, f(x^*) = x', f(x')$
24:         **end if**
25:         Update state-action value $Q(s, a)$
26:         Update new state $s \leftarrow s'$
27:     **end for**
28: **end while**
29: **Return:** Q-table

3.4 Determining the Order of the Local Search Operators in GRASP

We introduce an approach for optimizing the sequence of local search operators in the GRASP algorithm. Our approach leverages the concept of state-action value. The state-action value, often denoted as $Q(s, a)$, indicates the anticipated cumulative reward when an agent executes a specific action a while situated in a given state s, and subsequently follows a designated policy to maximize rewards from that point onward. To elaborate, if the state captures the previous action executed by the agent and the actions denote heuristic operators, then $Q(s, a)$ will embody the projected cumulative reward obtained by the agent when applying a specific local search N_{t+1} subsequent to another N_t. Following agent training, we can utilize the Q-table to determine the order of local search operators. This strategic arrangement holds the potential to enhance both solution quality and computational efficiency within the GRASP-PR algorithm.

The procedure for selecting the local search operators, namely GRASP-Q, is presented in Algorithm 4. The first step in GRASP at each iteration is to create a feasible solution for the problem. Therefore, the algorithm initializes the first state as $s = 0$, indicating that the previous action taken was 0 (Line 3). Based on the state s, the algorithm selects the next action a^*, corresponding to the operator following action 0, where the state-action value $Q(s, a)$ is the maximum (Line 5). The selected action a^* is added to the set of ordered operators O, and the state-action values $Q(s, a^*)$ for all states s are removed from the Q-table. This process is iterated until all the operators are ordered.

Algorithm 4. Selecting order of local search operators

1: **Input:** $Q - table$
2: Set of ordered operators, O
3: $s = 0$
4: **while** $|O| < 12$ **do**
5:     $a^* = \arg\max_a Q(s, a)$
6:     Add action a^* to O
7:     Remove all $Q(s, a^*)$ for all $s \in S$
8: **end while**
9: **Return:** Set of ordered operators, O

3.5 Q-Learning Based Hyperheuristic (HH-Q)

In the previous research, there have been efforts to utilize tabular-based methods like Q-learning and Sarsa to control actions in the search process for solving the VRP [14]. However, when using the previous taken action as a feature of the state in tabular-based methods, forming a deterministic policy by greedily selecting the action with the maximum state-action value is not effective. Following a greedy policy over the Q-table in these cases can lead the agent to become stuck in a cycle involving only two actions. For instance, if the previous action taken is $a_{t-1} = 0$, then the state $s_t = 0$, and the maximum Q-value at state $s_t = 0$ is $Q(0, 2)$, which implies the agent should take action $a_t = 2$ at step t. Subsequently, at step $t + 1$, the state s_{t+1} becomes 2 because the previous action taken was 2. Now, the maximum Q-value at state $s_{t+1} = 2$ is $Q(2, 0)$. This cycle continues iteratively throughout the time horizon.

We propose an alternative method to formulate a stochastic policy to be implemented based on the Q-values of a trained Q-table, rather than using a greedy approach. Instead of selecting a single action for a given state deterministically, we create a discrete distribution over actions from the Q-table for each state and sample an action from this distribution. This idea aligns with effective reinforcement learning algorithms like REINFORCE, Actor-critic, and Trust Region Policy Optimization (TRPO), where the policy is formed in a stochastic manner [19–21]. To begin, we normalize the trained Q-table by the following equation:

$$Q(s_i, a_i) = Q(s_i, a_i) - \min_{s_i \in S, a_i \in A} Q(s_i, a_i) \tag{4}$$

where $Q(s_i, a_i)$ represents the state-action value when taking action a_i in state s_i. After normalization, we create a discrete distribution over actions for a given state using Roulette Wheel selection, as shown in the equation below:

$$\pi(a_i|s_i) = \frac{Q(s_i, a_i)}{\sum\limits_{a_i \in A} Q(s_i, a_i)} \tag{5}$$

where $\pi(a_i|s_i)$ is the probability of taking action a_i at state s_i.

After constructing the discrete probability distribution over actions for all states from the Q-table, during the search, sampling an action from this distribution based on the current state creates an adaptive operator selection framework where the action with the highest Q-value will have the highest probability of being selected. Our policy encourages the agent to prioritize actions that can yield the maximum state-action value while also allowing for opportunities to take other actions that have the potential to lead to better solutions in a dynamic environment like HVRP-MTWSTT. This approach helps avoid getting stuck in a situation where only two actions are repeatedly chosen, as discussed earlier. We refer to this method as a Q-learning-based Hyperheuristic (HH-Q).

The pseudocode for deploying HH-Q to solve HVRP-MTWSTT is presented in Algorithm 5. HH-Q takes a Q-table trained using the training algorithm proposed in Sect. 3.3 as an input (Line 1). The search process begins by creating a feasible solution x for the problem and initializing the best solution x^* along with its objective value $f(x^*)$ (Lines 3–4). The initial state s_0 and solution pool P for the path-relinking procedure are also set up (Lines 4–5). The algorithm then takes actions for a maximum of i_{max} times to find a solution for the problem (Lines 7–21) iteratively. In each iteration, given the current state s_t, the action a_t is selected using the Roulette Wheel selection as defined in Eq. 5 (Line 8). The selected action a_t is then applied to the current solution to find a new solution x' for the problem (Lines 9–10). If the newly found solution satisfies the conditions to enter the elite solution pool P, it is added to P (Lines 11–12). If the selected action is 0, indicating the creation of a new solution, the search is re-initialized by setting the current solution to the new one (Lines 14–15). If the new solution is better than the current best solution, it becomes the best solution for the next iteration (Lines 17–18). At the end of each step, the new state s_{t+1} is updated for the next step (Line 20). After performing the action a maximum of i_{max} times, the algorithm returns the best solution found throughout the entire search process (Line 22).

4 Computational Experiments

4.1 Experiment Setting and Benchmark Instances

Benchmark Instances. The benchmark HVRP-MTWSTT instances are based on the deterministic versions [5]. To investigate the proposed methods with different sizes of instances, we selected three instances with varying numbers of nodes in the graph: 20, 50, and 100, respectively. Subsequently, we generated

Algorithm 5. Q-learning based Hyperheuristic

```
1: Input: trained Q-table;
2: Number of iteration i_{max};
3: Create a feasible solution x;
4: Initialize best solution x*, f(x*) ← x, f(x);
5: Initial random state, s_0;
6: Initialize pool P for path-relinking
7: for i = 0 to i_{max} do
8:      a_t ← select the action using Equation 5 given state s_t;
9:      Take action a_t;
10:     Observe new solution x';
11:     if f(x') > (1 − μ)f(x*) then
12:         Add x' to P;
13:     end if
14:     if action a_t create a new solution (a_t = 0) then
15:         x, f(x) ← x', f(x');
16:     end if
17:     if f(x') > f(x*) then
18:         x*, f(x*) ← x', f(x');
19:     end if
20:     Update new state s_{t+1};
21: end for
22: Return: x*, f(x*);
```

both training and testing instances for each size, resulting in a total of 200 instances for training and another 200 instances for testing purposes for each problem size. These instances were created by modifying the node coordinates and time windows compared to the original selected instances, while maintaining other specifications such as the number of vehicles and nodes consistent within each size. The characteristics of the training and testing instances are summarized in Table 1. N_1 and N_2 represent the number of on-demand requests and e-bins, whereas V_1 and V_2 indicate the number of vehicles of type I and type II, respectively. For vehicle type I, the associated parameters are a larger capacity of 500, operating cost of 2, penalty cost of 7, and idle cost of 3. For vehicle type II, the corresponding values are 100, 1, 2, and 5, respectively. A four-day collection period is considered in all benchmark instances.

Table 1. The train and test instances

Datasets	N_1	N_2	V_1	V_2	No. of instances
Instance20	10	10	5	5	400
Instance50	25	25	13	13	400
Instance100	50	50	25	25	400

We also prepare a set of instances with a variable number of nodes ranging from 8 to 110 for the scalability experiment (Table 2). The goal is to assess the efficacy of the proposed learning methods on unseen instances with varying numbers of nodes and graph structures, as opposed to changes in time windows and node coordination. The remaining parameters, such as capacity, operating costs, idle costs, penalty costs of vehicles, and the investigation length of the collection period, remain consistent with the other train and test instances.

Table 2. The problem instance characteristics for the scalability experiment

No	Instance	N_1	N_2	V_1	V_2
1	S-8	4	4	4	4
2	S-12	6	6	3	3
3	S-16	8	8	4	4
4	S-20	10	10	5	5
5	S-24	12	12	6	6
6	S-30	15	15	8	8
7	S-40	20	20	10	10
8	S-50	25	25	13	13
9	S-60	30	30	15	15
10	S-70	35	35	18	18
11	S-80	40	40	20	20
12	S-90	45	45	23	23
13	S-100	50	50	25	25
14	S-110	55	55	28	28

Experiment Settings and Baselines. All the experiments were conducted on a computer with 32 GB of RAM, an Intel(R) Core i5-12400 2.5 GHz CPU, and operating on a 64-bit Windows 10 Education system. The algorithms were implemented in Python 3.11. Turning to the Q-learning hyperparameters, the number of training episodes denoted as M, and the number of steps per episode denoted as m, are set to 500 and 6000, respectively (Lines 1 to 2 in Algorithm 3). The learning rate α of the Q-learning algorithm is assigned a value of 0.9, and the discount factor γ is set to 0.95 (Line 11 in Algorithm 2). In the Q-learning algorithm, actions are selected according to an ϵ-greedy policy (see Line 12 of Algorithm 3). The value of ϵ is initialized to 1.0 and is gradually decreased to a minimum of 0.05 during the training process, using a decay rate denoted as $\mathcal{D}$ = 0.0005. These parameters are set based on preliminary experiments.

To evaluate our proposed solutions, we use two baseline algorithms. The first algorithm is GRASP complemented with Path-relinking (GRASP-PR) [6]. We adapted the original GRASP-PR to handle the stochastic problem by utilizing the maximum distance to construct the initial solution for HVRP-MTWSTT, while keeping the other components of GRASP-PR similar to the original algorithm. The second baseline is SimGRASP [16], which was introduced for solving a VRP with stochastic demand. It employs Monte Carlo simulation to sample from the distribution of uncertainty. We integrated the concept of SimGRASP into the modified GRASP-PR to solve HVRP-MTWSTT. The parameters for SimGRASP were independently configured for each problem size, with the number of small runs and extended runs set to 30 and 150, 50 and 300, and 200 and 1000 for problem sizes of 20, 50, and 100, respectively. All the considered algorithms are terminated after 6000 applications of local search operators to enhance the initial solution.

4.2   Experimental Results

First, all methods solve the problem instances. The average and standard deviation (in bracket) of the best solutions found in these test instances and the average computational time per instance are presented in Table 3. The performance gaps in solution quality of GRASP-Q, SimGRASP, and HH-Q in comparison with GRASP-PR are illustrated in Fig. 1. It is observed that HH-Q performs the best among other algorithms for all test sets, except for the test set with a problem size of 20, where SimGRASP performs slightly better than HH-Q. For larger problem sizes of 50 and 100, the performance gap between HH-Q and the baseline algorithms becomes more pronounced. This observation can be attributed to the fact that for small-sized problem instances, the search space is not overly large, and the simulation method in SimGRASP can be effective in sampling from local optima to find good solutions. As the problem size increases and the search space expands, the simulation method shows its disadvantages when compared to HH-Q. GRASP-PR performs the worst among the considered methods, as it is primarily designed for deterministic VRP.

Table 3. Performance of proposed algorithm in solving 600 different instances

Datasets	GRASP-PR		GRASP-Q		SimGRASP		HH-Q	
	Profit	CPU(s)	Profit	CPU(s)	Profit	CPU(s)	Profit	CPU(s)
Instance20	751.01 (322.59)	6.03	802.81 (306.77)	6.77	**970.23 (284.87)**	8.13	940.11 (274.1)	**5.73**
Instance50	−2162.07 (1614.89)	21.66	−1935.23 (1606.53)	20.8	−1275.51 (1308.22)	26.05	**−830.49 (1350.35)**	18.09
Instance100	−20979.24 (6505.79)	82.7	−19562.53 (6149.17)	53.64	−18342.39 (6445.75)	91.48	**−14544.83 (5448.07)**	74.2

Additionally, GRASP-Q can only marginally improve GRASP-PR, with random order of operator, performance, achieving an improvement of ∼6–10%. This is expected as both GRASP-Q and GRASP-PR still use a fixed sequence order of operators, which demonstrates limitations in searching for good solutions in a stochastic VRP. Taking a closer look at the two algorithms, SimGRASP and HH-Q: in problem instance 20, SimGRASP outperforms HH-Q by approximately 2%, while for larger test sets, HH-Q clearly surpasses SimGRASP, with a performance gap of around 20%. This highlights that using an adaptive operator selection mechanism in solving HVRP-MTWSTT yields higher quality solutions compared to traditional metaheuristic methods.

Regarding the computational (CPU) times, it is worth noting that the time complexity of different operators varies, and the runtime of each operator heavily

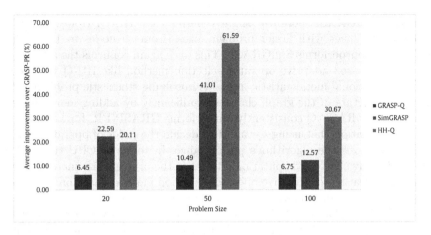

Fig. 1. Performance gap of all methods in comparison with GRASP-PR

depends on the specific solution to which an operator is applied. For GRASP-PR and GRASP-Q, since these algorithms use a fixed order of operators in each iteration, the CPU times are similar for instance sets with sizes of 20 and 50. However, for larger instances with 100 nodes, GRASP-Q demonstrates significantly shorter CPU times than GRASP-PR, averaging 53.64 s compared to 82.7 s. This illustrates the effectiveness of the method that employs Q-values to determine the order of operators within the GRASP framework (GRASP-Q) as the problem size increases. HH-Q exhibits the best performance in terms of CPU times in problem sizes 20 and 50. For problem size of 100, the CPU time of HH-Q is slightly shorter than that of GRASP-PR but longer than that of GRASP-Q. As for SimGRASP, due to the two simulation runs within the framework during the search, the CPU time for SimGRASP is significantly longer than for the other methods under consideration. The proposed methods based on Q-learning require time to learn a search policy from the instance set before implementation. Consequently, an additional computational effort is anticipated when applying these learning-based methods. However, the trade-off involving this extra computational effort during the learning phase can be justified when solving a set of problem instances with practical sizes.

For learning-based methods (e.g. GRASP-Q and HH-Q), generalization is of paramount importance in real-world applications. It refers to its ability to apply the knowledge acquired from train instances when handling new and unseen data. We conduct an experiment to evaluate the ability of the proposed methods in solving test instances (see Table 2) with different numbers of nodes and vehicles compared to the instances used to train GRASP-Q and HH-Q. We provide the average objective values from ten runs and the corresponding CPU times of the proposed methods in solving 14 different problem instances with varying numbers of nodes and vehicles in Table 4. We observe a similar trend as in the previous experiment on the test instances, where SimGRASP and HH-Q outperform GRASP-PR and GRASP-Q. For the small instances (from 8 to

24 nodes), SimGRASP performs slightly better than HH-Q. However, for the remaining instances with larger problem sizes (from 30 nodes to 110 nodes), HH-Q clearly outperforms SimGRASP. This once again confirms the advantages of a learning-based adaptive operator selection method like HH-Q when compared to traditional metaheuristic methods in solving stochastic problems, even when the structure of the graph changes significantly by adding more nodes. It is noted that GRASP-Q consistently outperforms GRASP-PR, further reinforcing the observation that using Q-values to decide the order of operators within GRASP improves the algorithm's performance. In terms of CPU time, HH-Q has significantly shorter computational times as the problem size increases, indicating that creating an adaptive operator selection method based on Q-learning offers favorable conditions for practical-sized problem instances.

Due to the stochastic nature of the algorithms, we verified the performance of HH-Q in comparison with reference algorithms using a non-parametric Friedman test, followed by a post-hoc test [22]. The findings indicate that HH-Q outperforms all the considered methods with $\alpha = 0.05$ for problem sizes of 50 and 100. For problem size of 20, SimGRASP exhibits a higher ranking than HH-Q, but the observed difference is not statistically significant.

Table 4. The scalability experiment results

No	Instance	GRASP-PR		GRASP-Q		SimGRASP		HH-Q	
		Avg. 10	CPU (s)	Avg. 10	CPU (s)	Avg. 10	CPU (s)	Avg. 10	CPU (s)
1	S-8	579.16	1.72	628.41	1.82	**680.06**	3.33	633.90	2.62
2	S-12	120.51	2.99	157.96	3.09	**325.07**	5.57	192.73	3.68
3	S-16	819.35	3.75	863.70	3.90	**1099.55**	6.35	856.17	4.00
4	S-20	542.06	5.12	540.31	5.47	**791.14**	8.59	729.77	5.53
5	S-24	2004.01	6.85	2000.28	6.82	**2171.34**	10.98	2098.93	6.39
6	S-30	619.73	10.12	718.93	9.96	770.70	19.96	**970.00**	8.83
7	S-40	−201.27	15.54	−295.29	13.08	409.94	20.14	**459.41**	11.56
8	S-50	−2613.90	23.86	−2529.79	17.39	−1065.21	25.16	**−1023.10**	15.37
9	S-60	−4270.13	36.23	−4795.87	30.31	−3682.71	33.56	**−1830.33**	20.81
10	S-70	−11886.78	45.68	−10980.31	42.97	**−8295.74**	40.91	−8702.46	24.16
11	S-80	−16401.37	50.97	−17159.66	42.96	−12024.53	59.31	**−10048.22**	32.98
12	S-90	−16077.14	64.81	−14949.01	48.77	−14958.22	85.76	**−9656.27**	41.88
13	S-100	−19455.57	72.75	−19064.94	53.75	−13907.32	91.72	**−13533.61**	47.02
14	S-110	−31065.17	75.52	−28593.94	85.30	−25965.34	107.73	**−22633.78**	58.30

5   Conclusions

We introduce two methods for integrating and utilizing Q-learning to solve the Heterogeneous VRP with Multiple Time Windows and Stochastic Travel Time problem. The first method uses the trained Q-table to determine the order of local search operators of GRASP. The second method introduces a hyperheuristic that employs a stochastic policy to select operators during the search process based on Q-values. Both are compared against the traditional metaheuristic GRASP-PR and SimGRASP. The experiments on test instances of three different problem sizes demonstrate that HH-Q outperforms all considered methods in medium problem sizes, while GRASP-Q modestly improves upon GRASP-PR. Furthermore, an experiment on the generalization of the trained methods with different instances shows that HH-Q performs better than all considered methods for problem sizes equal to or larger than 30 nodes, both in terms of solution quality and computational time.

These observations highlight the promise of integrating learning methods to create an adaptive operator selection framework for solving rich VRP instances. However, the current framework uses a single state, which has limitations in extracting information generated during the search procedure. For future work, advanced learning techniques like Deep Reinforcement Learning could be employed to handle states with more features, which has the potential to improve the algorithm's performance in solving the stochastic variant of VRP. Real world problems can be further considered to evaluate the performance of algorithms.

Acknowledgement. This research was supported by the Singapore Ministry of Education (MOE) Academic Research Fund (AcRF) Tier 1 grant.

References

1. Pérez-Belis, V., Bovea, M.D., Ibáñez-Forés, V.: An in-depth literature review of the waste electrical and electronic equipment context: trends and evolution. Waste Manage. Res. **33**(1), 3–29 (2015)
2. Wu, H., Tao, F., Yang, B.: Optimization of vehicle routing for waste collection and transportation. Int. J. Environ. Res. Public Health **17**(14), 4963 (2020)
3. Szwarc, K., Nowakowski, P., Boryczka, U.: An evolutionary approach to the vehicle route planning in e-waste mobile collection on demand. Soft. Comput. **25**(8), 6665–6680 (2021)
4. Pourhejazy, P., Zhang, D., Zhu, Q., Wei, F., Song, S.: Integrated e-waste transportation using capacitated general routing problem with time-window. Transp. Res. Part E: Logist. Transp. Rev. **145**, 102169 (2021)
5. Gunawan, A., Nguyen, M.P.K., Vincent, F.Y., Nguyen, D.V.A.: The heterogeneous vehicle routing problem with multiple time windows for the e-waste collection problem. In: 19th International Conference on Automation Science and Engineering (CASE) (2023)
6. Gunawan, A., Nguyen, D.V.A., Nguyen, P.K.M., Vansteenwegen, P.: Grasp solution approach for the e-waste collection problem. In: Daduna, J.R., Liedtke, G., Shi, X., Voß, S. (eds.) ICCL 2023. LNCS, vol. 14239, pp. 260–275. Springer, Cham (2023). https://doi.org/10.1007/978-3-031-43612-3_16
7. Król, A., Nowakowski, P., Mrówczyńska, B.: How to improve WEEE management? Novel approach in mobile collection with application of artificial intelligence. Waste Manage. **50**, 222–233 (2016)

8. Karimi-Mamaghan, M., Mohammadi, M., Meyer, P., Karimi-Mamaghan, A.M., Talbi, E.G.: Machine learning at the service of meta-heuristics for solving combinatorial optimization problems: a state-of-the-art. Eur. J. Oper. Res. **296**(2), 393–422 (2022)
9. Talbi, E.G.: Machine learning into metaheuristics: a survey and taxonomy. ACM Comput. Surv. (CSUR) **54**(6), 1–32 (2021)
10. Sutton, R.S., Barto, A.G.: Reinforcement Learning: An Introduction. MIT Press, Cambridge (2018)
11. Lu, H., Zhang, X., Yang, S.: A learning-based iterative method for solving vehicle routing problems. In: International Conference on Learning Representations (2019)
12. Peng, B., Zhang, Y., Gajpal, Y., Chen, X.: A memetic algorithm for the green vehicle routing problem. Sustainability **11**(21), 6055 (2019)
13. Reijnen, R., Zhang, Y., Lau, H.C., Bukhsh, Z.: Operator selection in adaptive large neighborhood search using deep reinforcement learning. arXiv preprint arXiv:2211.00759 (2022)
14. Ödling, D.: A metaheuristic for vehicle routing problems based on reinforcement learning (2018)
15. Watkins, C.J., Dayan, P.: Q-learning. Mach. Learn. **8**, 279–292 (1992)
16. Festa, P., Pastore, T., Ferone, D., Juan, A.A., Bayliss, C.: Integrating biased-randomized GRASP with monte carlo simulation for solving the vehicle routing problem with stochastic demands. In: 2018 Winter Simulation Conference (WSC), pp. 2989–3000. IEEE (2018)
17. Golden, B., Assad, A., Levy, L., Gheysens, F.: The fleet size and mix vehicle routing problem. Comput. Oper. Res. **11**(1), 49–66 (1984)
18. Kallestad, J., Hasibi, R., Hemmati, A., Sörensen, K.: A general deep reinforcement learning hyperheuristic framework for solving combinatorial optimization problems. Eur. J. Oper. Res. **309**(1), 446–468 (2023)
19. Sutton, R.S., McAllester, D., Singh, S., Mansour, Y.: Policy gradient methods for reinforcement learning with function approximation. In: Advances in Neural Information Processing Systems, vol. 12 (1999)
20. Konda, V., Tsitsiklis, J.: Actor-critic algorithms. In: Advances in Neural Information Processing Systems, vol. 12 (1999)
21. Schulman, J., Levine, S., Abbeel, P., Jordan, M., Moritz, P.: Trust region policy optimization. In: International Conference on Machine Learning, pp. 1889–1897. PMLR (2015)
22. Sheskin, D.J.: Handbook of Parametric and Nonparametric Statistical Procedures. CRC Press, Boca Raton (2020)

A Memetic Algorithm with Adaptive Operator Selection for Graph Coloring

Cyril Grelier⬤, Olivier Goudet⬤, and Jin-Kao Hao$^{(\boxtimes)}$⬤

LERIA, Université d'Angers, 2 Boulevard Lavoisier, 49045 Angers, France
{cyril.grelier,olivier.goudet,jin-kao.hao}@univ-angers.fr

Abstract. We present a memetic algorithm with adaptive operator selection for k-coloring and weighted vertex coloring. Our method uses online selection to adaptively determine the couple of crossover and local search operators to apply during the search to improve the efficiency of the algorithm. This leads to better results than without the operator selection and allows us to find a new coloring with 404 colors for C2000.9, one of the largest and densest instances of the classical DIMACS coloring benchmarks. The proposed method also finds three new best solutions for the weighted vertex coloring problem. We investigate the impacts of the different algorithmic variants on both problems.

Keywords: Graph Coloring · Memetic Algorithm · Hyperheuristics

1 Introduction

Graph coloring problems find applications across various domains, such as matrix decomposition [25], metropolitan area network design [14], and task scheduling in distributed computing [17]. Given a graph $G = (V, E)$, defined by its vertex set V and edge set E, a *legal* coloring S of the graph G partitions the vertex set V into k non-empty and disjoint color groups $\{V_1, \ldots, V_k\}$ such that the coloring constraint is satisfied, i.e., for each V_i, if $x \in V_i$ and $y \in V_i$, then $\{x, y\} \notin E$. In other words, the coloring constraint states that two adjacent vertices cannot go to the same color group (they cannot receive the same color). A coloring failing to meet the coloring constraint is an illegal coloring. Typically, graph coloring problems entail finding a legal coloring of the graph G while taking into account additional decision criteria and constraints. Specifically, the k-coloring problem (k-col) is a decision problem, where given a number of colors k, the goal is to find a legal coloring of the graph using these k colors. The graph coloring problem (GCP) is to determine the smallest number of colors (chromatic number of the graph) needed to color a graph. In the weighted vertex coloring problem (WVCP), an additional weight function $w : V \to \mathbb{R}^+$ is defined to assign a strictly positive weight $w(v)$ to each vertex $v \in V$. The objective of the WVCP is then to find a legal coloring $S = \{V_1, \ldots, V_k\}$ with a minimal score $f(S) = \sum_{i=1}^{k} \max_{v \in V_i} w(v)$, where the number of used colors k is not specified.

© The Author(s), under exclusive license to Springer Nature Switzerland AG 2024
T. Stützle and M. Wagner (Eds.): EvoCOP 2024, LNCS 14632, pp. 65–80, 2024.
https://doi.org/10.1007/978-3-031-57712-3_5

This paper aims to develop a learning-based framework for solving both the k-col decision problem and the WVCP minimization problem, which have been addressed by various methods in the literature. Both problems are known to be NP-hard in the general case [8], posing computational challenges in practice. For example, some random graphs with 250 vertices cannot be solved optimally by current exact algorithms (see [19] for k-col and [11] for WVCP). For this reason, a number of heuristics have been developed over the last thirty years to obtain approximate solutions to large graph coloring problems [6,20].

For the k-col, two particularly interesting local search heuristics are TabuCol [15] and PartialCol [2], which were proposed many years ago. For the WVCP, dedicated and effective local search procedures are much more recent, including AFISA [28], RedLS [29], ILS-TS [22] and TabuWeight [12]. None of these methods really dominates the others for all reference instances of the k-col and the WVCP [2,12,22]. It would be interesting to choose an appropriate local search to solve each type of instance for each problem.

However, local search methods may fail to produce high-quality solutions due to their limited ability to diversify their search. To overcome this difficulty, hybrid algorithms using the framework of memetic algorithms have been proposed, which benefit from local search for intensification and offer diversification possibilities with a population of high-quality solutions recombined with crossover operators. Hybrid algorithms have mainly been used to date to solve the k-coloring problem. The HEA (*Hybrid Evolutionary Algorithm*) algorithm [7] introduced the powerful GPX crossover (*Greedy Partition Crossover*) operator and used a local tabu search inspired by TabuCol. Evo-Div [23] and MACOL [18] both used crossover strategies with multiple parents and distance management between solutions. More recently, the HEAD algorithm (*HEA in Duet*) [21] proposes the use of only two individuals in the population and a reintegration system for high-quality individuals (elites) found earlier in the search. HEAD also uses the GPX crossover and an improved TabuCol algorithm. This algorithm HEAD is currently one of the most efficient solvers for the k-col. For the WVCP, to our knowledge, there is only one memetic algorithm in the literature, DLMCOL [10], which uses a large population (more than 20,000) and parallel GPU-based local searches, combined with a neural network-guided crossover selection.

It was observed by the authors of HEAD [21], that the quality (number of conflicts) of an offspring solution for the k-col, generated with the GPX crossover from two parents, is highly correlated with the partition distance [24] between these two parents, which is the minimum number of vertices that must change color to transform one solution to another. Based on this insight, the authors suggested employing modified versions of the GPX crossover, such as conservative asymmetric crossovers, which can lead to better results for certain types of instances. In the context of a memetic algorithm, it is therefore important to choose not only the right local search, but also the right crossover operator to perform an efficient search.

To this end, in this work, we investigate the use of online hyperheuristics, suggested in the literature to dynamically select adapted low-level heuristic components during the search process for solving a specific problem instance. We refer the reader to [3, 4] for an overview of existing hyperheuristics used to solve various combinatorial optimization problems. More specifically, hyperheuristics have been used to solve partitioning problems, with applications to planning and graph coloring [5, 13, 26, 27]. However, to our knowledge, no hyperheuristic-based memetic algorithm for graph coloring has yet been proposed in the literature. This work fills this gap by:

- investigating new memetic algorithms for the k-col and the WVCP using the HEAD framework [21] with diverse local search procedures and GPX variants.
- examining the ability of adaptive operator selectors to jointly choose crossovers and local search procedures during the search for specific instances.

In the rest of the paper, we first present our new general framework, AHEAD (for Adaptive HEAD), with different strategies for selecting local search procedures and crossovers (Sect. 2). Next, we show the results of the different variants of AHEAD in comparison with state-of-the-art algorithms (Sect. 3).

2   Adaptive Memetic Algorithm

In this section, we present the general framework of the adaptive memetic algorithm developed for the k-col and the WVCP, as well as the selection operators.

For the k-col, with a graph $G = (V, E)$ and k colors, the search explores the space Ω_k of legal and illegal colorings where all vertices are colored, but allowing color conflicts between adjacent vertices:

$$\Omega_k = \{\{V_1, \ldots, V_k\} : (\cup_{i=1}^{k} V_i = V) \wedge (V_i \cap V_j = \emptyset, i \neq j, 1 \leq i, j \leq k)\}. \quad (1)$$

In this case, the fitness f (to be minimized) corresponds to the number of conflicts in the solution. A solution with the fitness of 0 is a legal coloring.

For the WVCP, the algorithm works in the space of legal solutions Ω_l, where all vertices are colored with no limit on the number of colors, but no adjacent vertices are allowed to share the same color:

$$\Omega_l = \{\{V_1, \ldots, V_k\} : (\cup_{i=1}^{k} V_i = V) \wedge (V_i \cap V_j = \emptyset, i \neq j, 1 \leq i, j \leq k)$$
$$\wedge (\forall v_1, v_2 \in V_i, (v_1, v_2) \notin E, 1 \leq i \leq k)\} \quad (2)$$

and we search in this search space a solution $S = \{V_1, \ldots, V_k\}$ whose score $f(S) = \sum_{i=1}^{k} \max_{v \in V_i} w(v)$ is minimum.

2.1   Main Scheme

The architecture of the AHEAD framework, illustrated in Fig. 1, extends the state-of-the-art HEAD framework [21] by introducing an operator selector. In particular, its simplicity with only two individuals in the population facilitates the parent matching and population updating phases compared to other memetic algorithms in the literature [7,10,18,23].

AHEAD takes as input a graph $G = (V, E)$, and an integer value k for the k-col or a weight function w for the WVCP, a set of local search operators $\mathcal{O}^l$, a set of crossover operators $\mathcal{O}^x$, and a high-level selection strategy π_θ characterized by a parameter vector θ. The π_θ strategy chooses two pairs of operators $<o^l, o^x>$ with $o^l \in \mathcal{O}^l$ and $o^x \in \mathcal{O}^x$ to apply in the current generation to create two new individuals, replacing their parents in the population for the next generation.

The population is initialized with two random colorings, S_1 and S_2. Two elite solutions, E_1 and E_2, are also created, which are updated by the best solution found during the search and are used to reintroduce diversity into the population under specific conditions. Then, each generation of the algorithm performs the next six steps until the stopping condition is met.

1. A selection phase to select two pairs of operators $<$crossover, local search$>$ to be applied during this generation: $<o_1^x, o_1^l>$ and $<o_2^x, o_2^l>$. This selection is made using the function π_θ, which takes as input the two individuals in the population, with S_1 as the first input for the selection of $<o_1^x, o_1^l>$, then S_2 as the first input for the selection of $<o_2^x, o_2^l>$ (Sect. 2.2).
2. A crossover phase to create two offspring individuals C_1 and C_2 with $C_1 = o_1^x(S_1, S_2)$ and $C_2 = o_2^x(S_2, S_1)$ (see Sect. 2.3).
3. An intensification phase to obtain improved offspring solutions with local search, $C_1' = o_1^l(C_1)$ and $C_2' = o_2^l(C_2)$ (Sect. 2.4).
4. An insertion phase to replace S_1 and S_2 by C_1' and C_2', regardless of the fitness of the offspring solutions compared to the parents.
5. An update phase to adjust the π_θ selection policy from the examples collected over the last few generations (Sect. 2.5).
6. As in the original HEAD algorithm [21], each generation ends with a step of storing the best individual (*elite*) from the cycle (10 generations). The elite individual is reintroduced two cycles later (Sect. 2.6).

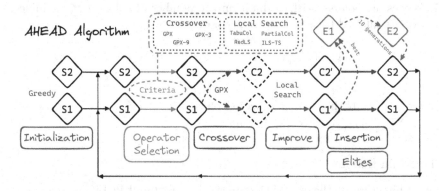

Fig. 1. General architecture of the Adaptive HEAD (AHEAD) framework.

2.2 Automatic Operator Selection

To select the crossover and local search operators, a high-level π_θ strategy automatically selects a pair of operators $<o^x, o^l> \in \mathcal{O}^x \times \mathcal{O}^l$. Therefore, there are $|\mathcal{O}^x| \times |\mathcal{O}^l|$ different <crossover, local search> pairs possible considered as independent meta-operators. The set of crossover operators $\mathcal{O}^x$ is presented in Sect. 2.3 and the set of local search operators $\mathcal{O}^l$ in Sect. 2.4.

In the general case, the π_θ function takes a pair of parents as input. For example, for (S_1, S_2), π_θ chooses a pair of operators $<o^x, o^l>$, with a crossover operator o^x to be applied with S_1 as the first parent and S_2 as the second parent. It will produce a child C_1, which will be improved by the local search procedure o^l to obtain a new individual C_1'.

In this work, we examine the effects of six operator selection policies π_θ with varying levels of complexity, including four fitness-based criteria, a neural network, and a random selector. Note that, except for the Deleter criteria, they are also used in combination with a Monte Carlo tree search in [13].

- *Random* performs a uniform selection among all operators.
- *Deleter* deletes the operator with the worst average result every 5 generations, until only one remains.
- *Roulette* (or Adaptive roulette wheel) [9] selects randomly the operator with a bias induced by the reward r (see Sect. 2.5) obtained by each operator. The better the reward is, the higher its associated probability of being picked is, using the same parameters as in [13].
- *UCB* (Upper Confidence Bound, One-armed bandit strategy) [1] selects the operator depending on the rewards obtained in previous generations and the number of times it has been picked using the UCB formula managing the exploitation-exploration trade-off.
- *Pursuit* [9] randomly selects the operator with a proportional bias in favor of the best performing operator. This strategy is among the most elitist, as it gives more chances to the best strategy applied in previous iterations.
- *NN* (Neural Network) uses the recommendations of a *deep set* neural network architecture [10,13,31]. This neural network g_θ takes as input a coloring S as a set of k binary vectors $\mathbf{v_j}$ of size n, $S = \{\mathbf{v_1}, \ldots, \mathbf{v_k}\}$, where each $\mathbf{v_j}$ indicates the vertices belonging to the color group j. From such an entry the neural network outputs a vector with $|\mathcal{O}^l|$ values in $\mathbb{R}$ corresponding to the expected reward of each local search. In order to select a pair of operators $<o^x, o^l> \in |\mathcal{O}^x| \times |\mathcal{O}^l|$, from a pair of parents (S_1, S_2), the neural network selector π_θ works as follows:
  1. Each crossover operator $o^x \in \mathcal{O}^x$ creates an individual $C^x = o^x(S_1, S_2)$.
  2. Each raw solution C^x is passed as input to the neural network g_θ to obtain a vector of $|\mathcal{O}^l|$ values in $\mathbb{R}$ corresponding to the estimated score that can be obtained after applying each local search $o^l \in \mathcal{O}$ to C^x.
  3. The pair $<o^x, o^l>$ corresponding to the highest output value of the neural network (for the $|\mathcal{O}^x|$ evaluations given by g_θ) is selected. 10% of the time, the selection is random to encourage diversity.

2.3   Application of Crossover Operators

In this step, the parent solutions S_1 and S_2 are combined to create two offspring solutions $C_1 = o_1^x(S_1, S_2)$ and $C_2 = o_2^x(S_2, S_1)$ from the crossover operators o_1^x and o_2^x chosen during the selection phase detailed in the previous section.

For both problems, we use the popular GPX crossover [7] used in the original HEAD algorithm [21]. It consists of alternately taking the largest color group from each parent and transmitting it to the offspring solution. Note that GPX is asymmetrical, applying it with the pair (S_1, S_2) or the pair (S_2, S_1) does not produce the same offspring solution. Three variants of this crossover can be selected in AHEAD: GPX, GPX-3, and GPX-9 taking respectively 1, 3, and 9 groups of color in the first parent for 1 color in the second parent. Therefore the last two are more conservative than the original GPX as more groups of the first parent are transmitted to the offspring. This generally results in offspring with lower (better) fitness, but less different from its parents. The two new solutions C_1 and C_2 generated during the crossover phase will be improved in the local search phase detailed in the next subsection.

2.4   Local Searches

In this step, the new individuals C_1 and C_2 are improved by the selected local search operators o_1^l and o_2^l. These two independent local searches are run in parallel on two CPUs with a time limit of T_{LS} seconds. The best solutions C_1' and C_2' found by o_1^l and o_2^l with this time budget are returned. For the k-col, two state-of-the-art local search operators can be chosen: the efficient implementation of TabuCol [15] proposed in [21] and PartialCol [2]. For the WVCP, the two best performing algorithms RedLS [29] and ILS-TS [22] can be chosen.

2.5   Operator Selection Strategy Update

At each generation, new learning examples are collected from the results of pairs of operators to update the selection strategy for future generations.

Learning Examples Memory. Reward scores $r_1 = -f(C_1')$ and $r_2 = -f(C_2')$ are associated with the choice of operator pairs $<o_1^x, o_1^l>$ and $<o_2^x, o_2^l>$, with f the fitness function of the k-col or the WVCP. These rewards are negative, as the fitness f is to be minimized in both problems. Then, two learning examples (o_1^x, o_1^l, C_1, r_1) and (o_2^x, o_2^l, C_2, r_2) are stored in a database D, specific to the current execution. D is a queue of the N last examples obtained during previous generations (N is set to 50 empirically). This limited queue size enables us to better adapt to potential variations in operator results, in the event that certain operators are better at the beginning of the search than at the end.

Online Learning of the Selection Criteria. Every generation, the π_θ policy is trained on the database D and all its θ parameters are updated. For the NN policy, the training phase occurs every $nb = 20$ generations. During this training phase, each training example (o^x, o^l, C, r) from the database D is converted into a supervised learning example (X, y), with X an input matrix of size $k \times |V|$ corresponding to the set of k vectors $C = \{\mathbf{v_1}, \ldots, \mathbf{v_k}\}$, and y is a real vector of size $|\mathcal{O}^l|$ (number of local search operators), so that y is initialized with $g_\theta(C)$, the output vector of the neural network taking as input C, then its value $y[o^l]$ for the chosen operator o^l is replaced by the expected reward r: $y[o^l] = r$. Once this conversion of the training examples is done, g_θ is trained to minimize the mean square error computed over the $|\mathcal{O}^l|$ outputs (supervised learning) on this training dataset for 10 epochs with the Adam optimizer [16].

2.6 Insertion and Elite Solutions

Like HEAD [21], offspring solutions systematically replace the parents and elite solutions are used. Elites, which are high-quality solutions from previous generations, are added to the population after 10 generations. Thus, every 10 generations, the best individual encountered from the previous 11 to 20 generations is reintegrated into the population.

Furthermore, at each generation, the set-theoretic partition distance [24] between the solutions S_1 and S_2 is evaluated. This distance is defined as the least number of one-move operator changes for transforming S_1 to S_2. If the distance is 0, meaning that the two individuals are the same solution, the individuals of the population are randomly reinitialized. Note that when this happens, we do not reset the θ weights of the learning strategy in order to benefit from what the operator selector has learned since the start of the search.

3 Computational Experiments

This section is dedicated to a computational assessment of the proposed AHEAD algorithm for solving the k-coloring and the weighted vertex coloring problems, by making comparisons with state-of-the-art methods. We also discuss the impact of the different operator selection strategies on the results.

3.1 Experimental Settings

In these experiments, we consider 31 instances for the k-col and 48 instances for the WVCP, among the most challenging DIMACS benchmark instances and widely used in the experiments of many recent papers [21,22,28]. 20 independent runs per method and per instance are carried out for one hour with two CPU for the HEAD and AHEAD methods (four hours for the WVCP) and two hours with one CPU for the local search algorithms for the k-col (eight hours for the WVCP). For each k-col instance, and for each independent run, the smallest value of k for which the method is able to find a legal solution is reported.

The time spent, in seconds, in the local search during each generation in HEAD and AHEAD is $0.001 * |V|$ for the k-col and $0.04 * |V|$ for the WVCP, with $|V|$ being the number of vertices in the graph. These values were chosen following tests on a range of values for each problem, which will not be presented here. To erase the impact of the training time of the neural network, for the versions with AHEAD, the methods perform as many generations in the memetic algorithm as the HEAD versions.

The experiments are run on a computer equipped with an Intel Xeon ES 2630, 2.66 GHz processor. All algorithms are coded in C++, compiled and optimized with the g++ 12.1 compiler. For the neural network implementation, the Pytorch 1.13 C++ library was used. The source code and complete tables of detailed results are available at https://github.com/Cyril-Grelier/gcp_ahead and https://github.com/Cyril-Grelier/wvcp_ahead.

3.2   Experimental Results for the k-Coloring Problem

In this section, we first analyze the general results obtained for the k-coloring problem. The different methods tested can be regrouped into three categories:

- The standalone local searches PartialCol (PC) [2] and TabuCol (TC) [15] (with optimizations from [21]).
- The memetic algorithm HEAD [21] combined with local search operators PartialCol (version HEAD + PC) and TabuCol (version HEAD + TC).
- The different proposed AHEAD versions with the 6 different operator selection strategies presented in Sect. 2.2. These versions are called AHEAD + the name of the selection strategy.

We first present a general comparison between all these methods, followed by detailed results on the different benchmark instances.

General Comparisons. Table 1 displays general performance comparisons between each pair of methods on the benchmark instances considered in this work for the k-col. Whenever the mean score of a method in the row for an instance is better than the mean score of a method in the column, and this difference is significant (non-parametric Wilcoxon signed rank test with p-value ≤ 0.001) the method in the row obtains one point. If the method in the row is better on more instances than the method in the column than the opposite, then the number of instances is in bold. As an example, we observe in Table 1 that TabuCol is significantly better on 14 instances when compared with PartialCol, while PartialCol is only better on 2 instances when compared with TabuCol. The last three columns of Table 1 gives the number of times a Best Known Score (BKS) from the literature is found by the method and the number of times the method reaches the best score and the best mean among the presented methods.

We observe in Table 1 that using the memetic framework HEAD with PartialCol or TabuCol (version HEAD + TC, and HEAD + PC) improves the results over the methods using the corresponding local search alone.

Overall, TabuCol is more effective than PartialCol, that is why HEAD+TC keeps good results against some versions of AHEAD with less elitist operator selection such as Random, Roulette or UCB. However, the other versions of

Table 1. Comparison of each method for the k-col, the value is the number of instances where the method in the row is significantly better than the method in the column. The last three columns are a summary of the number of best scores.

/31 Instances	PartialCol	TabuCol	HEAD+PC	HEAD+TC	AHEAD+Random	AHEAD+Roulette	AHEAD+Deleter	AHEAD+UCB	AHEAD+Pursuit	AHEAD+NN	# BKS	# Best Score	# Best Mean
PartialCol	-	2	3	2	1	2	1	2	2	2	5	8	11
TabuCol	14	-	11	2	2	1	0	2	0	1	8	14	7
HEAD+PC	8	6	-	1	0	0	1	0	0	0	6	10	7
HEAD+TC	18	12	20	-	4	2	1	2	2	2	7	17	15
AHEAD+Random	17	11	19	1	-	0	1	1	0	0	9	17	9
AHEAD+Roulette	17	11	19	1	0	-	0	0	0	0	11	19	12
AHEAD+Deleter	19	15	20	5	8	3	-	5	1	1	13	24	20
AHEAD+UCB	19	11	20	1	1	0	0	-	0	0	10	18	10
AHEAD+Pursuit	19	13	20	3	5	2	0	1	-	0	11	20	14
AHEAD+NN	19	12	20	2	4	0	0	0	0	-	12	23	16

AHEAD, using Deleter, Pursuit and the neural network (NN), obtained overall better results than the memetic algorithm HEAD+TC without operator selection. It highlights the interest of dynamically choosing the best operator to apply for each given instance.

The Random selection policy is less effective in comparison with the other operator selection strategies, especially against Deleter, Pursuit and NN, which have a stronger bias on selecting the best operators. Surprisingly, the simplest but most elitist selection strategy, Deleter, achieves the best results, indicating that for this problem, once the best operator has been identified for each specific instance, there is generally no need to change it for the rest of the search.

Detailed Results. Table 2 shows, for each instance, the Best Known Score (BKS) in the literature[1] with a star if it is optimal, then, for each method, the best score, the mean score and the average time to reach the best scores over the 20 executions. Bold values indicate the best scores among the studied methods. The average score is not shown if equal to the best score. Due to space limitations, only a selection of methods is shown in the various tables, and not all instances studied are shown. Complete tables are available on the github repository.

First, those results confirm that the TabuCol local search is more often better than PartialCol, but that the latter can give better results for some

---

[1] Achieving these BKS for k-col, especially for the largest instances, is a very difficult task. Some have only been found by few algorithms under particular conditions (hyperparameter tuning, extended execution times of several days to a month).

Table 2. Results of the main methods for the GCP.

instance	BKS	PartialCol			TabuCol			HEAD+TC			AHEAD+Random			AHEAD+Deleter		
		best	mean	time	best	mean	time	best	mean	time	best	mean	time	best	mean	time
C2000.5	145	164	165.2	5313	162	162.8	4628	**148**	149.2	3330	150	150.7	3101	149	150.7	3152
C2000.9	408	420	420.8	5171	411	412.5	4786	**405**	406.4	2328	**405**	407.7	2956	**404**	**405.6**	2988
C4000.5	259	304	305.6	6690	303	304.2	5567	**278**	279.6	3580	280	281.6	3651	279	280.8	3404
DSJC500.1	12	**12**		128	**12**		75	**12**		86	**12**		80	**12**		56
DSJC500.5	47	50	50.1	2227	49		460	**48**		819	**48**		1258	**48**		850
DSJC500.9	126	128		975	**126**	126.3	2988	**126**		1027	**126**	126.1	1379	**126**		632
DSJC1000.1	20	21		1	21		0	21		0	21		1	**20**	20.9	2391
DSJC1000.5	82	90	90.5	3516	88		1760	**83**	83.3	2290	**83**	83.5	2372	**83**	83.5	2511
DSJC1000.9	222	227	228.4	3630	224	224.9	3345	**223**	224	1616	**223**	224.2	2734	**223**	223.8	1589
DSJR500.5	122*	125	126.2	1666	124	127	1155	**123**	124	1766	**123**	124.2	2245	**123**	123.8	2289
flat300_28_0	28*	**28**		896	**28**	29.5	3220	30	30.8	1916	**28**	28.5	702	**28**	30.4	5
flat1000_50_0	50*	**50**		44	**50**		69	**50**		28	**50**		8	**50**		8
flat1000_60_0	60*	**60**		213	**60**		233	**60**		54	**60**		28	**60**		29
flat1000_76_0	76*	89	89.1	2845	86	87	3096	**82**	82.3	1905	**82**	82.8	2775	**82**	82.8	1969
latin_square_10	97	107	110.2	4875	100	100.8	4377	102	103.7	93	103	103.8	1996	**99**	100.7	1729
le450_25c	25*	27		69	26		0	26		0	**25**	25.9	1407	**25**	25.3	1022
le450_25d	25*	27		50	26		0	26		0	26		0	**25**	25.3	1537
queen11_11	11*	**11**	**11.9**	1303	12		0	12		0	12		0	12		0
queen12_12	12*	13		4	13		0	13		0	13		0	13		1
queen13_13	13*	14		20	14		0	14		1	14		1	14		1
queen14_14	14*	15		585	15		20	15		9	15		18	15		16
r250.5	65*	67		134	66	67.2	462	**65**	66	3378	**65**	66	1638	66		549
r1000.1c	98	141	149.1	61	134	155.2	77	**100**	101.6	264	**100**	101.6	1674	**100**	101.6	1621
r1000.5	234	247	248.1	5638	**244**	245.6	3622	246	247.6	1479	246	247.4	2134	245	**245.5**	2009
wap01a	41*	42		1088	42	43	2160	42		137	42		143	**41**	**42**	1958
wap02a	40*	41	41.7	4275	**40**	41.1	6499	41		15	41		15	**40**	40.8	1634
wap03a	43	44		91	44	45.9	4342	45		261	45		87	**43**	44.3	2387
wap04a	41	43		61	**42**	43.1	4869	43		880	43		1186	43		293
wap06a	40*	41		98	**40**	41.3	4248	**40**		909	**40**	40.8	1549	**40**		246
wap07a	41	44		41	**41**	42.3	5046	42	42.1	1771	42	43	2526	42	42.1	494
wap08a	40*	43	43.2	2750	**41**	**41.5**	2967	42		48	42		365	**41**	41.9	2146
#BKS		5/31			8/31			7/31			9/31			13/31		
#Best		8/31			14/31			17/31			17/31			24/31		
#Best Avg		11/31			7/31			15/31			9/31			20/31		

instances such as queen11_11 and can be faster for solving other instances such as flat300_28_0 or wap03a. It shows to some extent that these two local searches can be complementary.

Second, we obviously confirm that when TabuCol is integrated within the HEAD memetic framework and combined with the GPX crossover (version HEAD +TC), it can generally improve the results significantly, but it does not improve the results for all instances. For example, for the instances flat300_28_0, r1000.5, and some wap, it is actually better to use the local search alone for better intensification. Using crossovers for these instances can actually disrupt the search too early, preventing the local search from significantly improving its results. Note that the results of HEAD in [21] can differ from our results with HEAD+TabuCol using exactly the same operators, because we did not perform a fine-tuning of the number of iterations spent in local search for each given instance, unlike it was done in the original article.

Third, the algorithm AHEAD + Random can find the best results for more instances than the HEAD+TabuCol version whose choice of operators does not change during the search. For example, AHEAD + Random finds the optimal coloring with 28 colors for the instance flat300_28_0. This is due to the fact that TabuCol is not able to reach the chromatic number of the graph in a short amount time compared to PartialCol which can reach it systematically more than three times faster. Therefore, AHEAD + Random benefits from having a chance at each generation to select the right local search for each given instance.

Fourth, from Table 2, we observe that using an elitist strategy in local search and crossover selection can significantly improve the results compared to the random selection strategy. In particular, as shown in this table, the version AHEAD + Deleter can find a new best coloring with $k = 404$ for the instance C2000.9 that has never been reported in the literature. This new best score is also found by the versions AHEAD + Pursuit and AHEAD + NN. In general, these versions of AHEAD with elitist operator selection strategies obtain the best results for a wide variety of instances of different types.

Figure 2 shows the average cumulative selections for each pair of operators, performed by the different selection criteria for the DSJC500.1 and queen12_12 instances. In the plots, TabuCol and PartialCol selections are indicated by red and blue lines, respectively, and a higher contrast indicates a more conservative crossover. When we look at the frequencies of the local search operators selected by AHEAD, we see that TabuCol is selected most often in comparison with PartialCol, which is no surprise, as TabuCol is already better on its own for a greater number of instances. However, when it comes to crossovers, we observe a balanced choice between the three GPX variants, with a bias toward the more conservative crossover GPX-9 for geometric graphs (e.g., DSJR500.5) and sparse graphs (e.g., wap instances), for which local optima are very distant in the search space, while the GPX crossover is more often preferred for random and dense graphs (e.g., DSJC1000.9), for which there is often larger backbones of solutions shared by the high-quality solutions (as shown in [10]).

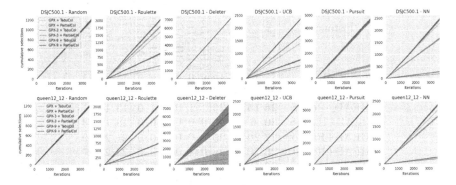

Fig. 2. Average cumulative selections, along with error bars, for each pair of operators based on different criteria on the DSJC500.1 and queen12_12 instances.

3.3 Experimental Results for WVCP

Now, we analyze the general results obtained for the WVCP. We first present a general comparison between the methods, followed by detailed results on the different benchmark instances. The studied methods are the following:

- The local searches, RedLS [29] and ILS-TS [22] (8h runs on 1 CPU).
- The memetic algorithm HEAD [21] with RedLS or ILS-TS (HEAD + RedLS /ILS-TS) and the crossover GPX (4h runs on 2 CPUs).
- The different proposed AHEAD versions with the 6 different operator selection strategies presented in Sect. 2.2 with the two local searches (RedLS and ILS-TS) and the three variants of GPX crossovers (4h runs on 2 CPUs).

General Comparisons. As seen in Table 3, using HEAD with RedLS (HEAD + RedLS) improves the results for 26 instances but the standalone local search RedLS stays better in 9 instances. On the other hand, for ILS-TS, using the HEAD framework is better only for 6 instances, while ILS-TS remains better for 19 instances. We observe that the use of crossovers in combination with the ILS-TS local search procedure does not improve the results of ILS-TS. This can be explained by the fact that ILS-TS is already a method incorporating a strong perturbation strategy for search diversification, making the use of crossovers somewhat superfluous. Regarding the results of AHEAD, we confirm what we have observed for the k-col, even if it's less pronounced. The AHEAD versions are more often better than the other methods, and the most elitist operator selection strategies (Deleter and Pursuit) obtain the best results.

Table 3. Comparison of each method for the WVCP, the value is the number of instances where the method in the row is significantly better than the method in the column.

/48 Instances	RedLS	ILS-TS	HEAD+RedLS	HEAD+ILS-TS	AHEAD+Random	AHEAD+Roulette	AHEAD+Deleter	AHEAD+UCB	AHEAD+Pursuit	AHEAD+NN	# BKS	# Best Score	# Best Mean
RedLS	-	10	9	14	9	9	9	9	9	9	15	24	11
ILS-TS	27	-	8	19	3	6	5	4	1	3	23	25	21
HEAD+RedLS	26	15	-	25	1	1	0	0	1	0	19	19	11
HEAD+ILS-TS	20	6	5	-	0	0	0	0	0	0	18	19	13
AHEAD+Random	27	20	10	25	-	0	0	0	0	0	21	22	19
AHEAD+Roulette	26	20	9	26	0	-	0	0	0	0	22	22	17
AHEAD+Deleter	26	19	9	26	3	0	-	0	0	0	**24**	**28**	19
AHEAD+UCB	26	20	9	26	1	1	0	-	0	0	23	23	19
AHEAD+Pursuit	26	23	11	26	1	0	0	0	-	0	**24**	26	**22**
AHEAD+NN	27	21	10	27	0	1	0	0	0	-	21	23	19

Table 4. Results of the best methods for the WVCP.

instance	BKS	RedLS			ILS-TS			HEAD+RedLS			AHEAD+Random			AHEAD+Deleter		
		best	mean	time	best	mean	time	best	mean	time	best	mean	time	best	mean	time
C2000.5	2144	**2131**	**2155.7**	18367	2244	2264.4	6423	2244	2257.9	7453	2220	2236.8	12962	2218	2236.3	1782
C2000.9	5477	**5439**	5455.1	23137	5847	5910.1	23014	5732	5748.2	12980	5732	5783.9	12491	5717	5758.8	12327
DSJC1000.1	300	303	306.9	5839	305	306.2	5819	304	305.6	7380	302	303.8	9348	**300**	**302.2**	12874
DSJC1000.5	1185	**1190**	**1206.9**	12204	1241	1267.7	21935	1225	1229.7	7011	1222	1228.2	5371	1224	1230.5	1476
DSJC1000.9	2836	**2828**	**2841.8**	22796	3004	3035.9	25345	2909	2926.5	820	2911	2928.7	12633	2907	2926.8	2379
DSJC500.1	184	187	194	702	185	187.3	7107	186	186.9	6594	185	186.5	10290	**184**	185.9	8022
DSJC500.5	685	707	712.5	27147	711	721.2	9150	709	712.6	2534	**706**	**711.5**	12516	709	713.5	5838
DSJC500.9	1662	**1667**	**1671**	9925	1709	1725.3	24351	1680	1683.5	4053	1678	1684.2	12644	1676	1682.8	8149
DSJC250.1	127	129	131.4	56	**127**	127.1	11901	**127**		4516	**127**		3729	**127**	127.2	3235
DSJC250.5	392	399	400.8	2602	**392**	**393.9**	10722	395	396.2	8349	393	395.2	9592	**392**	396.6	6028
DSJC250.9	934*	934	935	9679	934	935.1	14740	934	935.1	6741	**934**	**934.2**	8097	934	935	5011
DSJC125.5gb	240	243	252.7	0	**240**		132	**240**	240.9	4098	**240**		222	**240**		152
DSJC125.5g	71	72		1063	**71**		64	**71**		1609	**71**		86	**71**		104
DSJC125.9gb	604*	**604**		2	**604**		125	**604**		4	**604**		13	**604**		12
DSJC125.9g	169*	**169**		0	**169**		320	**169**		0	**169**		6	**169**		9
flat1000_50_0	924	**1152**	**1165.7**	6259	1213	1230.5	570	1181	1187.7	7544	1179	1186.3	4428	1180	1186.8	2952
flat1000_60_0	1162	**1196**	**1204.8**	1877	1247	1263.8	25765	1216	1227.2	10824	1213	1223.7	11726	1217	1224.5	9840
flat1000_76_0	1165	**1163**	**1183.2**	28084	1228	1242.2	16513	1192	1204	2214	1187	1203	10742	1196	1204	8938
latin_square_10	1480	**1505**	**1515.3**	14189	1555	1575	18924	1523	1532.5	11286	1510	1526.2	13987	1517	1527.8	8732
le450_15a	212	213	215.4	54	**211**	213.6	11684	**212**	212.8	6777	**212**	212.8	8819	**211**	**212.4**	10557
le450_15b	216	218	219.9	41	217	217.1	10346	**216**	217	3204	**216**	217.1	2736	**215**	**216.5**	11124
le450_15c	275	282	285.4	82	279	281.7	16288	277	279.4	8360	277	**278.8**	7220	278	279.4	4788
le450_15d	272	277	280.6	325	275	277.6	8456	274	276.1	6004	274	275.6	8759	**273**	275.2	13299
le450_25a	306	**306**	306.6	2881	**306**		142	**306**		161	**306**		169	**306**		131
le450_25b	307*	**307**	307.6	95	**307**		23	**307**		53	**307**		28	**307**		19
le450_25c	342	348	352.8	583	348	349.1	16413	347	348.1	180	**346**	**347.8**	5652	**346**	348	588
le450_25d	330	335	339.4	232	337	338.7	14212	**333**	334.4	5904	**333**	334.2	6282	**333**	334.2	9648
queen10_10	162	162	164.8	865	**162**		20	**162**		51	**162**		32	**162**		27
queen10_10gb	164	165	168.7	4790	**164**		172	**164**	164.4	4850	**164**		227	**164**		314
queen10_10g	43*	**43**	43.1	12	**43**		7	**43**		11	**43**		7	**43**		9
queen11_11	172	174	178	28766	**172**		6983	**172**	172.7	5668	**172**	172.1	3108	**172**		2207
queen11_11gb	176	177	178.6	1329	**176**		187	**176**		1583	**176**		436	**176**		396
queen11_11g	47	**47**	47.9	669	**47**		154	**47**		276	**47**		144	**47**		151
queen12_12	185	188	189.9	61	**185**	**185.2**	13770	186	186.3	6201	185	185.6	5997	185	185.4	7066
queen12_12gb	191	192	197.8	150	**191**		5019	**191**	191.3	5174	**191**		1521	**191**	191.1	3380
queen12_12g	50	**50**	51.5	986	**50**		1214	**50**		1464	**50**		1533	**50**		865
queen13_13	194	**194**	199.9	8	**194**	194.8	11188	**194**	194.2	5243	**194**		1560	**194**	194.1	1480
queen14_14	215	218	223.8	568	**215**	216.4	9862	216	216.6	7956	**215**	216.2	6384	**214**	**215.3**	8624
queen15_15	223	228	229.7	5806	225	226.5	15730	**224**	225.5	13260	**224**	225.1	10180	**224**	225.7	7200
queen16_16	234	237	240.8	17	237	238.3		235	236.4	913	236	235	5610	235	236.4	11836
wap01a	545	557	577	995	**547**	550.1	20531	552	559.1	8178	549	553.6	14094	549	552.8	8874
wap02a	538	554	572.1	16183	**536**	541	21912	550	557.1	13884	541	546.1	7654	541	545.5	12994
wap03a	562	**569**	575.5	17878	572	575.5	22637	577	579.7	6992	573	576.3	8096	573	575.9	2944
wap04a	563	**567**	578.9	13939	**567**	570.5	7346	573	575.6	3152	570	573.2	1970	569	572.5	13790
wap05a	541	**542**	543.8	7719	**542**	542.2	11809	**542**	542.9	4471	**542**	543	12056	**542**	543.2	2772
wap06a	516	519	526.1	1575	**516**	519.5	6264	519	520.7	12180	518	521	9100	520	521.2	5978
wap07a	555	**554**	573	8460	565	569.2	16299	557	559.4	3360	558	559.8	12040	557	559.2	12460
wap08a	529	**536**	543.7	19557	543	546.9	19271	539	540.8	7452	539	541.2	1800	538	**540.1**	10608
#BKS		15/48			23/48			19/48			21/48			24/48		
#Best		24/48			25/48			19/48			22/48			28/48		
#Best Avg		11/48			21/48			11/48			19/48			19/48		

Detailed Results. First, we see in Table 4 that with the help of the eight hours of computation, RedLS is capable of finding five new scores (underlined score). ILS-TS is also able to find two new upper bounds with this longer execution time.

The RedLS local search alone remains better on large instances (e.g., C2000, latin_square and flat) than the different memetic versions using this local search procedure (HEAD + RedLS, and all AHEAD variants). Contrary to what was observed for the k-col, this shows that for very large WVCP instances, using the GPX crossover is not very beneficial. This can be explained by the fact that for the WVCP, only the maximum weight of each color group affects the score. Thus, for large instances, many different groupings of vertices are possible without impacting the score, which generally results in a very high distance between the two solutions S_1 and S_2 of the population. However, as observed in [21], the solution quality of an offspring built with the GPX crossover is poorer (higher fitness) if the individuals are too distant in the search space.

However, for medium-sized instances, such as le450_15a/b and queen14_14, the AHEAD memetic framework with an elitist operator selection (AHEAD + Deleter) significantly improves results and yields three new best upper bounds.

Regarding the operators selected by AHEAD, the two local search operators RedLS and ILS-TS are almost equally preferred, with a choice depending on the type of instance. On the other hand, unlike the k-col, the choice of crossover is almost exclusively oriented towards the most conservative crossover, GPX-9, particularly in combination with the local search ILS-TS. As mentioned above, this is due to the large distance between individuals in the population in the case of the WVCP.

4   Conclusion

The proposed AHEAD (Adaptive HEAD) framework is based on a population of two individuals and uses learning-driven operator selectors to determine a pair of local search and crossover to apply during the search process for solving a given instance of the k-coloring and weight vertex coloring problems. For both problems, the proposed approach shows advantages over versions without automatic selection of low-level operators. In the course of these experiments, we obtained three new best scores for the WVCP with the proposed AHEAD method, as well as a new best coloring with 404 colors for the very large and dense graph C2000.9.

The work could be extended by considering a wider variety of complementary crossover procedures and local searches to be chosen by the high level operator selection strategy. Future work could also involve coupling the choice of operators with the setting of critical hyperparameters involved in these operators, such as the number of local search iterations to be performed at each generation, or the size of the tabu list.

Acknowledgment. We would like to thank Dr. Yiyuan Wang, [30] and Pr. Bruno Nogueira [22] for sharing their codes. This work was granted access to the HPC resources of IDRIS (Grant No. AD010611887R1) from GENCI and the Centre Régional de Calcul Intensif des Pays de la Loire (CCIPL). We are grateful to the reviewers for their comments.

References

1. Auer, P.: Using confidence bounds for exploitation-exploration trade-offs. J. Mach. Learn. Res. **3**(Nov), 397–422 (2002)
2. Blöchliger, I., Zufferey, N.: A graph coloring heuristic using partial solutions and a reactive tabu scheme. Comput. Oper. Res. **35**(3), 960–975 (2008)
3. Burke, E.K., Gendreau, M., Hyde, M., Kendall, G., Ochoa, G., Özcan, E., Qu, R.: Hyper-heuristics: a survey of the state of the art. J. Oper. Res. Soc. **64**(12), 1695–1724 (2013)
4. Drake, J.H., Kheiri, A., Özcan, E., Burke, E.K.: Recent advances in selection hyper-heuristics. Eur. J. Oper. Res. **285**(2), 405–428 (2020)
5. Elhag, A., Özcan, E.: A grouping hyper-heuristic framework: application on graph colouring. Expert Syst. Appl. **42**(13), 5491–5507 (2015)
6. Galinier, P., Hamiez, J.P., Hao, J.K., Porumbel, D.: Recent advances in graph vertex coloring. In: Zelinka, I., Snášel, V., Abraham, A. (eds.) Handbook of Optimization. Intelligent Systems Reference Library, vol. 38, pp. 505–528. Springer, Heidelberg (2013). https://doi.org/10.1007/978-3-642-30504-7_20
7. Galinier, P., Hao, J.K.: Hybrid evolutionary algorithms for graph coloring. J. Comb. Optim. **3**, 379–397 (1999)
8. Garey, M.R.: Computers and Intractability: A Guide to the Theory of NP-Completeness. Freeman. Fundamental (1997)
9. Goëffon, A., Lardeux, F., Saubion, F.: Simulating non-stationary operators in search algorithms. Appl. Soft Comput. **38**, 257–268 (2016)
10. Goudet, O., Grelier, C., Hao, J.K.: A deep learning guided memetic framework for graph coloring problems. Knowl.-Based Syst. **258**, 109986 (2022)
11. Goudet, O., Grelier, C., Lesaint, D.: New bounds and constraint programming models for the weighted vertex coloring problem. In: Proceedings of the Thirty-Second International Joint Conference on Artificial Intelligence, IJCAI 2023, 19th–25th August 2023, Macao, SAR, China, pp. 1927–1934 (2023)
12. Grelier, C., Goudet, O., Hao, J.-K.: On Monte Carlo tree search for weighted vertex coloring. In: Pérez Cáceres, L., Verel, S. (eds.) EvoCOP 2022. LNCS, vol. 13222, pp. 1–16. Springer, Cham (2022). https://doi.org/10.1007/978-3-031-04148-8_1
13. Grelier, C., Goudet, O., Hao, J.K.: Monte Carlo tree search with adaptive simulation: a case study on weighted vertex coloring. In: Pérez Cáceres, L., Stützle, T. (eds.) EvoCOP 2023. LNCS, vol. 13987, pp. 98–113. Springer, Cham (2023). https://doi.org/10.1007/978-3-031-30035-6_7
14. Halldórsson, M.M., Shachnai, H.: Batch coloring flat graphs and thin. In: Gudmundsson, J. (ed.) SWAT 2008. LNCS, vol. 5124, pp. 198–209. Springer, Heidelberg (2008). https://doi.org/10.1007/978-3-540-69903-3_19
15. Hertz, A., Werra, D.D.: Using tabu search techniques for graph coloring. Computing **39**(4), 345–351 (1987)
16. Kingma, D.P., Ba, J.: Adam: a method for stochastic optimization. arXiv preprint arXiv:1412.6980 (2014)
17. Liu, H., Beck, M., Huang, J.: Dynamic co-scheduling of distributed computation and replication. In: Sixth IEEE International Symposium on Cluster Computing and the Grid (CCGRID 2006), vol. 1, pp. 9–pp. IEEE (2006)
18. Lü, Z., Hao, J.K.: A memetic algorithm for graph coloring. Eur. J. Oper. Res. **203**(1), 241–250 (2010)
19. Malaguti, E., Monaci, M., Toth, P.: An exact approach for the vertex coloring problem. Discret. Optim. **8**(2), 174–190 (2011)

20. Malaguti, E., Toth, P.: A survey on vertex coloring problems. Int. Trans. Oper. Res. **17**(1), 1–34 (2010)
21. Moalic, L., Gondran, A.: Variations on memetic algorithms for graph coloring problems. J. Heuristics **24**, 1–24 (2018)
22. Nogueira, B., Tavares, E., Maciel, P.: Iterated local search with tabu search for the weighted vertex coloring problem. Comput. Oper. Res. **125**, 105087 (2021)
23. Porumbel, D.C., Hao, J.K., Kuntz, P.: An evolutionary approach with diversity guarantee and well-informed grouping recombination for graph coloring. Comput. Oper. Res. **37**(10), 1822–1832 (2010)
24. Porumbel, D.C., Hao, J.K., Kuntz, P.: An efficient algorithm for computing the distance between close partitions. Discret. Appl. Math. **159**(1), 53–59 (2011)
25. Prais, M., Ribeiro, C.C.: Reactive grasp: an application to a matrix decomposition problem in TDMA traffic assignment. INFORMS J. Comput. **12**(3), 164–176 (2000)
26. Sabar, N.R., Ayob, M., Qu, R., Kendall, G.: A graph coloring constructive hyper-heuristic for examination timetabling problems. Appl. Intell. **37**(1), 1–11 (2012)
27. Sghir, I., Hao, J.-K., Ben Jaafar, I., Ghédira, K.: A distributed hybrid algorithm for the graph coloring problem. In: Bonnevay, S., Legrand, P., Monmarché, N., Lutton, E., Schoenauer, M. (eds.) EA 2015. LNCS, vol. 9554, pp. 205–218. Springer, Cham (2016). https://doi.org/10.1007/978-3-319-31471-6_16
28. Sun, W., Hao, J.K., Lai, X., Wu, Q.: Adaptive feasible and infeasible tabu search for weighted vertex coloring. Inf. Sci. **466**, 203–219 (2018)
29. Wang, Y., Cai, S., Pan, S., Li, X., Yin, M.: Reduction and local search for weighted graph coloring problem. In: Proceedings of the AAAI Conference on Artificial Intelligence, vol. 34, pp. 2433–2441 (2020)
30. Wang, Y., Cai, S., Pan, S., Li, X., Yin, M.: Reduction and local search for weighted graph coloring problem. In: Proceedings of the AAAI Conference on Artificial Intelligence, vol. 34, no. 0303, pp. 2433–2441 (2020)
31. Zaheer, M., Kottur, S., Ravanbakhsh, S., Póczos, B., Salakhutdinov, R., Smola, A.J.: Deep sets. In: Guyon, I., et al. (eds.) Advances in Neural Information Processing Systems 30: Annual Conference on Neural Information Processing Systems 2017, 4–9 December 2017, Long Beach, CA, USA, pp. 3391–3401 (2017)

Studies on Multi-objective Role Mining
in ERP Systems

Simon Anderer[1]([✉]), Bernd Scheuermann[2], and Sanaz Mostaghim[3]

[1] SIVIS GmbH, Grünhutstraße 6, 76187 Karlsruhe, Germany
Simon.Anderer@sivis.com
[2] Karlsruhe University of Applied Sciences, 76133 Karlsruhe, Germany
Bernd.Scheuermann@h-ka.de
[3] Otto-von-Guericke-Universität Magdeburg, Magdeburg, Germany
Sanaz.Mostaghim@ovgu.de

Abstract. A common concept to ensure the security of IT systems, in which multiple users share access to common resources, is Role Based Access Control (RBAC). Permissions, which correspond to the authorization to perform an operation on a data or business object are grouped into roles. These roles are then assigned to users. The corresponding optimization problem, the so-called Role Mining Problem (RMP), aims at finding a role concept comprising a minimal set of such roles and was shown to be NP-complete. However, in real-world role mining scenarios, it is typically the case that, besides the number of roles, further key figures must be consulted in order to adequately evaluate role concepts. Therefore, in this paper, the RMP is extended to a multi-objective (MO) optimization problem. Potential optimization objectives are discussed in the context of Enterprise Resource Planning (ERP) systems. Furthermore, it is shown, how evolutionary algorithms for the RMP can be adapted to meet the requirements of MO role mining. Based on this, the integration of different optimization objectives is examined and evaluated in a series of experiments.

Keywords: Role based access control · Enterprise resource planning systems · Evolutionary algorithms · Multi-objective optimization

1 Introduction

IT systems of companies and organizations must not only be protected against external attacks such as malware and phishing, but must also include concepts to prevent erroneous or fraudulent behavior of internal actors. The recent edition of the PricewaterhouseCoopers *Global Economic Crime and Fraud Survey* reports that almost half of the over 1,200 participants experienced cases of fraud. Of this, 31% were attributable to employees and another 26% were caused by collusions of internal and external perpetrators [1]. One approach to address this is to limit the access of users to sensitive data using access control mechanisms. A widely used concept is Role Based Access Control (RBAC), where permissions are grouped into roles, which are then assigned to users [2]. The corresponding NP-complete

© The Author(s), under exclusive license to Springer Nature Switzerland AG 2024
T. Stützle and M. Wagner (Eds.): EvoCOP 2024, LNCS 14632, pp. 81–96, 2024.
https://doi.org/10.1007/978-3-031-57712-3_6

optimization problem is called the *Role Mining Problem* (RMP) and aims at finding a role concept comprising a minimal set of roles and an assignment of these roles to users [3]. Enterprise Resource Planning (ERP) systems, which are used to support the business processes of companies, frequently use RBAC to manage the access of users to data. In this context, however, it is not sufficient for a role concept to comprise as few roles as possible in order to be suitable for the application in practice. Other aspects that should be taken into account when evaluating the quality of a role concept for ERP systems are, for example, the associated license costs as well as its adherence to compliance rules. The RMP, which was previously mainly considered as single-objective optimization problem, must therefore be extended in such a way that multiple objectives can be considered simultaneously resulting in a multi-objective (MO) version of the RMP. One advantage of considering the RMP as an MO optimization problem is that a decision maker does not have to specify his preferences before the optimization, but can choose from a set of role concepts after the optimization.

2   The Role Mining Problem

In the following, the main elements of the RMP are introduced. Based on this, the RMP can be defined as matrix decomposition problem [3]. At this, $U = \{u_1, u_2, ..., u_M\}$ denotes a set of $M = |U|$ users, $P = \{p_1, p_2, ..., p_N\}$ a set of $N = |P|$ permissions and $R = \{r_1, r_2, ..., r_K\}$ a set of $K = |R|$ roles. Furthermmore, $UPA \in \{0,1\}^{M \times N}$ is the targeted permission-to-user assignment matrix, where $UPA_{i,j} = 1$ implies that permission p_j is assigned to user u_i. $UA \in \{0,1\}^{M \times K}$ is a role-to-user assignment matrix and $PA \in \{0,1\}^{K \times N}$ is a permission-to-role assignment matrix. Finally, $RUPA := UA \otimes PA \in \{0,1\}^{M \times N}$ is the resulting permission-to-user assignment matrix, where $(UA \otimes PA)_{i,j} = \max_{l \in \{1,...,k\}} (UA_{i,l} \cdot PA_{l,j})$ denotes the Boolean Matrix Multiplication.

Definition 1. *The Basic Role Mining Problem Given a set of users U, a set of permissions P and a permission-to-user assignment matrix UPA, find a minimal set of Roles R, a corresponding role-to-user assignment matrix UA and a permission-to-role assignment matrix PA, such that each user is assigned exactly the set of permissions granted by UPA:*

$$min \ |R|, \quad s.t. \quad d(UPA, RUPA) = 0, \tag{1}$$

where $d(A, B) := \|A - B\|$ denotes the distance of two matrices $A, B \in \mathbb{R}^{m \times n}$ and $\|A\|$ is the sum of absolute values of elements of A: $\|A\| := \sum_{i=1}^{m} \sum_{j=1}^{n} |A_{i,j}|$.

A role concept $\pi := \langle R, UA, PA \rangle$, consisting of a set of roles R, a role-to-user assignment UA and a permission-to-role assignment PA, denotes a candidate solution for a given Basic RMP. The set of all candidate solutions for the Basic RMP is denoted Π. A candidate solution is called a feasible solution, if it satisfies the constraint in (1). For the Basic RMP, in particular, a feasible solution is also denoted 0-consistent. In order to further reduce the number of roles obtained from solutions of the Basic RMP, there are other well-known variants of the RMP, in which deviations between the actual and the targeted permission-to-user assignment are permitted $(UPA \neq RUPA)$, such that the 0-consistency constraint in (1) is relaxed, like δ-approx. RMP, Min. Noise RMP and Edge RMP, see [3]. Figure 1 shows an exemplary 0-consistent role concept π_0. For better visualization, black cells indicate 1's and white cells represent 0's.

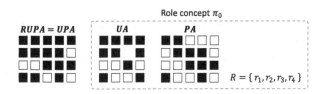

Fig. 1. Exemplary role concept π_0.

3 Solution Strategies for the RMP

Since the RMP is a well-known optimization problem, a variety of approaches for solving the RMP have been proposed. Many authors use permission grouping to create a set of candidate roles, which are then assigned to users mostly using greedy approaches [4–8]. Other authors map the RMP to different known optimization problems, e.g., the Set Cover Problem [9], the Minimum Tiling problem [3] or the Minimum Biclique Cover Problem [10], and use related solution strategies. Further approaches are based on formal concept analysis [23], graph optimization [11,12] or evolutionary algorithms [14,15]. A broad survey on different solution strategies for the RMP is provided by Mitra in [13]. In 2020, Anderer et al. presented the *addRole-EA* for the Basic RMP [16]. As this will serve as a basis for the adaption to MO role mining in the further course of this paper, it is presented in more detail at this point.

3.1 The addRole-EA

The addRole-EA is an evolutionary algorithm specifically designed for the Basic RMP. Except for an additional pre- and post-processing procedure, it comprises the common structure and components of an EA, see Fig. 2. In the following, the different components of the addRole-EA are briefly described, focusing on those relevant in the context of MO role mining. For a detailed description of the addRole-EA and its methods, refer to [16].

Encoding: Each individual of the addRole-EA represents one possible role concept. Therefore, its chromosome comprises a set of roles R, a role-to-user assignment matrix UA and a permission-to-role assignment matrix PA, as illustrated in Fig. 1.

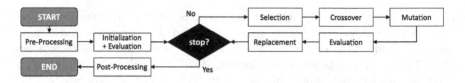

Fig. 2. Top-level description of addRole-EA, based on [16].

Pre- and Post-processing: In order to reduce problem size, four pre-processing steps have been identified comprising for example the removal of users that are assigned the same set of permissions except for one representative. The post-processing step adapts the obtained role concepts to the original problem size.

Initialization: Since the addRole-EA is designed for the Basic RMP, all role concepts need to satisfy the 0-consistency constraint. Therefore, each individual of the initial population is created in two steps. First the role-to-user assignment and the permission-to-role assignment matrix are preset as $UA := UPA$ and $PA := I_N$ (N-dimensional identity). Subsequently, a series of random mutation operators is carried out on each individual in order to increase diversity.

Selection, Crossover, Mutation and Replacement: The mutation operator of the addRole-EA creates new roles, which are then added to the chromosome of an individual. The crossover operator is used to exchange roles between individuals. This is possible without violating the 0-consistency constraint since, in case a role can be assigned to a user in one individual without deviation, it can also be assigned to the same user in every other individual. The individuals for mutation and crossover are selected based on a random selection scheme. Mutation as well as crossover are based on the addRole-method, which adds new roles to the chromosome of an individual and subsequently deletes all roles that have become obsolete in such a way that the 0-consistency constraint is not violated. Hence, only feasible individuals are obtained from this method. To determine the individuals of the next generation's population, a combination of elitism and random selection is used.

4 Objectives for Role Mining in ERP Systems

In this section, different possible optimization objectives are discussed. First, it is shown how tolerating deviations between UPA and $RUPA$ may lead to a decrease in the number of roles of a role concept. Subsequently, two optimization objectives are discussed which are of great relevance in the context of ERP systems: the *compliance score*, which measures the adherence of a role concept to compliance rules and the *license costs*, which measure the license costs associated with a role concept. To conclude this section, an overview of other optimization objectives that have been described in role mining literature is provided.

4.1 Deviations

Allowing for deviations between UPA and $RUPA$, may lead to role concepts that comprise less roles compared to role concepts that fulfill the 0-consistency constraint. In the following, this will be illustrated based on two examples: A natural approach to reduce the number of roles in π_0 (see Fig. 1) consists in merging two roles. In Fig. 3 (left), a new role r_{new} has been created from merging roles r_2 and r_3, resulting in role concept π_1. In this case, users u_2 and u_3 are each assigned an additional permission. Even though such *positive deviations* do not prevent users from performing their work, they can cause for violations of a company's compliance rules or an increase of license costs. However, especially if the assignment of permissions to users is derived from historical user behavior, it is often necessary to include positive deviations in order to assign all required permissions, since historical data usually does not cover the entire spectrum of tasks of a user [18]. Another straight-forward approach to reduce the number of roles consists, for example, in simply deleting one or more roles from a role concept. Figure 3 (right) shows role concept π_2, resulting from role concept π_0 after deletion of role r_1. It can be seen that the deletion of role r_1 only affects u_4 by the fact that he or she lacks permission p_2 and might thus not be capable of performing all tasks of his or her work. Such *negative deviations* should therefore be treated with caution and preferably be avoided. This is why only positive deviations are considered in the remainder of this paper.

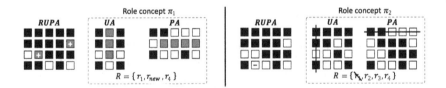

Fig. 3. Reducing number of roles by allowing for deviations.

4.2 Compliance Score

There are combinations of permissions that allow for malicious actions if assigned to the same user, so-called *Separation of Duties (SoD) conflicts*. For example, a user that is assigned the permission to edit the banking details in a vendor master record as well as the permission to execute payments is capable of temporarily exchanging the vendor's bank account for the own private bank account, execute the payment and undo the changes later, so as to disguise the malicious action. This is referred to as *vendor flipflop* and constitutes one of the standard examples for fraud committed by internal actors [19]. In order to include the consideration of SoD-conflicts into the role mining process, these can be aggregated into a compliance matrix $C \in \{0, 1\}^{L_C \times N}$. At this, L_C equals the total amount of SoD-conflicts and each row in C represents one SoD-conflict. Analogous to the permission-to-user assignment matrix UPA (or $RUPA$), each column corresponds to a permission. The permissions are to be arranged in the same order, such that $C_{l,j} = 1$ implies that the l-th SoD-conflict includes permission p_j. Comparing the compliance matrix C against the actual permissions assigned to a user as encoded in $RUPA$, one obtains, the matrix $\delta(\pi) \in \{0,1\}^{M \times L_C}$, representing the SoD-conflicts of each user, as:

$$\delta_{il}(\pi) := \begin{cases} 1, \text{if } \sum_{j=1}^{N} RUPA_{i,j} \cdot C_{l,j} = \sum_{j=1}^{N} C_{l,j} \\ 0, \text{else.} \end{cases} \tag{2}$$

To further differentiate the level of severeness of SoD-conflicts, an additional weight vector $w \in \mathbb{R}^{L_C}$ is introduced. Based on this, the so-called *compliance score* $CS(\pi)$ of a role concept π can be defined [17]:

$$CS(\pi) := \sum_{i=1}^{M} (\delta(\pi) \cdot w)_i. \tag{3}$$

It is clear that negative deviations have the potential to reduce the compliance score, whereas positive deviations may increase the compliance score. Further, the compliance score is constant and the same for all 0-consistent role concepts.

4.3   License Costs

In order to determine the license costs of a role concept, the named-user license model of SAP ERP is used. This means that an individual license must be purchased for each user of the ERP system which is dependent of the permissions assigned to the user. At this, each permission is assigned to a license category. Since there can be large price differences between different license categories, it may be worthwhile to reconsider the assignment of permissions to users minimizing license costs. For this, a permission-to-license-category assignment matrix $PL \in \{0,1\}^{N \times L}$, where L denotes the number of license categories and $PL_{i,j} = 1$, if p_i is assigned to license category L_j is introduced. Further, a license price vector $LP = (LP_1, ..., LP_L)^T$ assigns a price to each license category. Comparing these against the actual permissions assigned to a user as encoded in $RUPA$, one obtains the license costs $LC(\pi)$ of a role concept π, as follows:

$$LC(\pi) := \sum_{i=1}^{M} max_{j \in \{1,...,N\}} (RUPA_{i,j} \cdot (PL \cdot LP)_j). \tag{4}$$

Analogous to the analysis of the compliance score, it can be seen that negative deviations have the potential to reduce the license costs, whereas positive deviations may increase the license costs of a role concept. Again, license costs are constant and the same for all 0-consistent role concepts.

4.4   Further Optimization Objectives

In addition to license costs and compliance score, further criteria to assess the quality of role concepts have been proposed in literature. In order to support the decision on whether a new role concept should be deployed in a company, Saenko and Kotenko propose the consideration of the number of differences between a new role concept and the role concept currently deployed in a company [20]. Colantonio et al. [21] propose the inclusion of administrative costs to evaluate role concepts and consider the following weighted sum as the new objective function for the RMP $f(\pi) = \alpha\|UA\| + \beta\|PA\| + \gamma|R| + \delta\sum_{r\in R}c(r)$, where $\alpha, \beta, \gamma, \delta \geq 0$. At this, the administrative costs consist of the number of assignments of roles to users $\|UA\|$ and of permissions to roles $\|PA\|$ as well as the number of roles $|R|$. Furthermore, for each role r additional costs $c(r)$ related to other business information, for example, based on user attributes, are included. Depending on the values of α, β, γ and δ, different other variants originate from that approach, such as the *User-Oriented RMP*, in which the assignments of roles to users are included as additional optimization objective. Xu and Stoller consider the meaningfulness of roles. For this purpose, they include user attributes into the role mining process. Based on that, so-called *attribute expressions* are created from different values of the attributes, which aim at describing the semantic scope of a role. If a highly matching attribute expression is found for a role, this role is considered to be meaningful [22]. Molloy et al. also consider multiple objectives: They use formal concept analysis and user attributes to create meaningful roles [23,24]. Furthermore, they propose the *Weighted Structural Complexity*, which is also based on using a weighted-sum, for their evaluation. Stoller and Bui include these concepts into the framework of temporal role mining (TRBAC) [25]. One disadvantage of using a weighted sum approach is that it represents the a-priori preferences of a decision maker and cannot be adjusted thereafter. Particularly in real-world role mining scenarios, where the trade-offs between the different objectives are usually not clear in advance, considering the RMP as MO optimization problem can be advantageous, as it offers decision makers the possibility to decide whether a role concepts with fewer roles and thus less administrative effort, or a role concept with lower licensing costs or an improved adherence to compliance constraints would better fit the scenario under consideration after the optimization process is terminated.

5   Multi-objective Role Mining

In this section, a general definition of the multi-objective RMP (MO-RMP) is provided. For this purpose, a set $\Pi_F \subseteq \Pi$ is introduced, which denotes the set of all feasible role concepts. This can be flexibly adapted to the role mining scenario under consideration. In case there are no restrictions to a role concept, $\Pi_F = \Pi$. If only role concepts that comply with the 0-consistency constraint are to be considered, $\Pi_F = \{\pi \in \Pi : d(UPA, RUPA) = 0\}$. Based on this, the MO-RMP can be defined as follows:

Definition 2. *The Multi-objective RMP*

Given a set of users U, a set of permissions P and a targeted permission-to-user assignment matrix UPA, find role concepts $\pi = \langle R, UA, PA \rangle \in \Pi_F$, such that the objectives encoded by $f : \Pi_F \to \mathbb{R}^J$ are minimized:

$$min \ f_j(\pi), \quad j = 1, 2, ..., J, \quad s.t. \quad \pi \in \Pi_F. \tag{5}$$

This definition can be adapted flexibly to different role mining scenarios, with different optimization objectives, reflected in the multi-dimensional objective function f as well as different restrictions on the role concepts reflected in Π_F.

5.1　The Multi-objective addRole-EA

In the following, it is described how different components of the addRole-EA can be adapted to become applicable for the MO-RMP. The remaining components of the addRole-EA can be used in their original version as described in Sect. 3.

Adaption of Fitness Function: It is obvious that considering MO role mining scenarios requires an adaption of the hitherto one-dimensional fitness function of the addRole-EA. In the following section, two different role mining scenarios are considered which include a two-dimensional, a three-dimensional and a four-dimensional fitness function.

Adaption of Pre- and Post-processing: The deletion of permissions that are not assigned to any user as well as users that are not assigned any permission has no consequence in the context of MO role mining and could therefore be included into the MO-addRole-EA without modification. The aggregation of users into user classes can also be included into the MO-addRole-EA. However, in order to calculate the values of the different optimization objectives, such as the number of deviations, the compliance score or the license costs of a role concept, the cardinalities of the emerging user classes have to be taken into account. Considering the aggregation of permissions into permission classes, only those permissions can be aggregated into permission classes, which are assigned to exactly the same users, are contained in the same license category and additionally are included in the same SoD-conflicts. Therefore, the aggregation of permissions into permission classes is omitted within the MO-addRole-EA. In the post-processing procedure, the pre-processing steps included in the MO-addRole-EA are undone and the obtained role concepts are adapted to the initial role mining scenario.

Adaption of addRole-Method: As discussed in Sect. 4, within this paper, only positive deviations are permitted, resulting in $\Pi_F = \Pi^+$, where $\Pi^+ := \{\pi \in \Pi : (RUPA - UPA)_{i,j} \geq 0, \forall i, j\}$. To prevent roles from being arbitrarily assigned to users, it is checked whether a role and a user share at least one permission. Only if this is true, a role can be assigned to a user in the MO-addRole-EA. In addition, a parameter $d_{max}^+ \geq 0$ is introduced to limit the number of positive deviations for each user, comparing the actual $(RUPA)$ and the desired number of permissions (UPA) of user u_i: $\sum_{j=1}^{N} RUPA_{i,j} \leq (1 + d_{max}^+) \cdot \sum_{j=1}^{N} UPA_{i,j}$.

Adaption of Replacement: Comparing the fitness values of individuals in MO optimization problems must consider all fitness criteria. Therefore, the elitism approach used in the original version of the addRole-EA must be replaced. For this purpose the well-known *Non-Dominated Sorting Genetic Algorithm* (NSGA-II), which was introduced by Deb et al. in 2000 [26], is used as replacement.

6   Experiments and Evaluation

In this section, two different role mining scenarios are examined. However, the focus is not on detailed performance testing but rather on showing selected effects which seem relevant for the adaptation of the addRole-EA to MO role mining in the real-world application context. The different evaluation scenarios are based on common benchmark instances for role mining as well as corresponding extension files for the inclusion of compliance score and license costs, available within *RMPlib*, a public library for role mining benchmarks [17]. For evaluation, the benchmark instances *PLAIN_ small_ 02* (PS_02), containing 50 users and 50 permissions, as well as *PLAIN_ small_ 05* (PS_05), containing 100 users and 100 permissions, were selected. In order to include compliance score and license costs, the benchmark instances are combined with compliance respectively license costs extension files, which are also available within *RMPlib*, as follows: PS_02 is used together with compliance extension *CMPL_ 50_ 1.cmpl* and license costs extension *LIC_ 50_ 1.lic*. PS_05 is used in combination with *CMPL_ 100_ 1.cmpl* and *LIC_ 100_ 1.lic*. The two compliance extension files each contain 50 SoD-conflicts with different weights ranging from 0 to 20. The two license costs extension files each contain five license categories associated with prices between \$0 and \$6,000. As discussed in Sect. 4, compliance score and license costs are constant and the same for each 0-consistent role concept. For PS_02, the corresponding values $CS_0 = 1,876$ and $LC_0 = 225,800$. For PS_05, $CS_0 = 469$ and $LC_0 = 311,800$.

6.1   Scenario 1: Roles and Deviations

In the first evaluation scenario, additionally to the number of roles, the number of deviations is included into the role mining process as further optimization objective, such that $f : \Pi_F \rightarrow \mathbb{R}^2$, $f(\pi) = (|R|, \ d(RUPA, UPA))^T$.

To guarantee that each user is assigned at least all permissions needed to perform the tasks of his or her work, only positive deviations are allowed. Hence, $\Pi_F = \Pi^+$. At first, the influence of the upper bound d_{max}^+ is examined. For this purpose, the MO-addRole-EA was run 20 times on *PS_ 02* and *PS_ 05* with $d_{max}^+ \in \{0.5, 1.0, \infty\}$ with different random seeds. The values of the parameters of the addRole-EA are adopted from [16]. Figure 4 shows all non-dominated role concepts obtained from the final populations of all runs of the addRole-EA for the different values of d_{max}^+ after $t = 100,000$ iterations on PS_02. Since throughout Scenario 1, the same effects were obtained for PS_02 and PS_05, only the results obtained on PS_02 are shown.

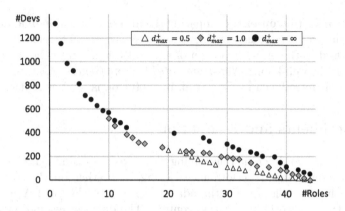

Fig. 4. Non-dominated role concepts for different values of d_{max}^+ on PS_02.

A first observation is that role concepts obtained from smaller values of d_{max}^+ seem to dominate the role concepts obtained from larger values of d_{max}^+ at least within the range of possible deviations restricted by the smaller value of d_{max}^+. However, larger values of d_{max}^+ allow for role concepts comprising less roles. It can be seen that in the case of unrestricted positive deviations $d_{max}^+ = \infty$, a role concept has been found involving only one role which is assigned all permissions and is assigned to all users, thus causing more than 1,300 positive deviations. Such solutions are mathematically possible, but unsuitable for the application in real-world scenarios, which endorses using d_{max}^+ as control parameter.

As indicated in [17], it is theoretically possible to obtain a 0-consistent role concept comprising at most 25 roles for PS_02. However, irrespective of the value of d_{max}^+, the MO-approach leads to 0-consistent role concepts that comprise more than 40 roles. In order to analyze this in more detail, Fig. 5 (left) shows the progression of the average number of roles over all role concepts and all runs on PS_02. Figure 5 (right) shows the average number of positive deviations.

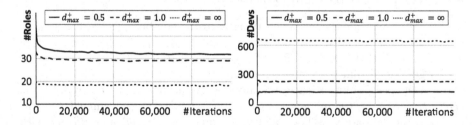

Fig. 5. Average number of roles/deviations on PS_02.

It can be seen that, although the initial individuals comprise no deviations, which is due to the initialization method of the addRole-EA, the number of deviations increases to its maximum level controlled by d_{max}^+ within a few iterations.

It is further shown that thereafter neither the number of roles nor the number of deviations can substantially be improved. A possible approach to address this is to initially consider the deviation-free Basic RMP as defined in Sect. 1 before including deviations as second objective. Figure 6 shows the corresponding results for $d_{max}^+ = 0.5$, where deviations were permitted either from the beginning at $t_1 = 0$ or after the execution of $t_2 = 25{,}000$ iterations on PS_02.

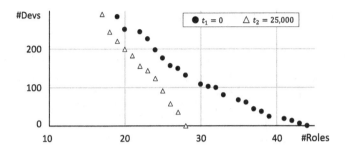

Fig. 6. Immediate (t_1) and delayed (t_2) admittance of deviations on PS_02.

Obviously, delayed admittance of deviations leads to 0-consistent role concepts with fewer roles compared to admitting them right from the beginning. Furthermore, all role concepts created using the delayed deviations approach clearly dominate the role concepts created with immediate deviations. This can possibly be explained by the fact that in case of delayed admittance of deviations, in the beginning of the optimization process, the structure of UPA is exploited to reduce the number of roles, whereas, in the case, where deviations are permitted from the beginning, the reduction of the roles is mainly achieved by including deviations. This is supported by Fig. 7 (left), which shows the progression of the average number of roles over iterations, and Fig. 7 (right), which shows the progression of the average number of roles over iterations, for $t_1 = 0$ and $t_2 = 25{,}000$, again considering $d_{max}^+ = 0.5$ on PS_02. Based on these findings, one option to possibly further improve the quality of the obtained role concepts could be to use d_{max}^+ as control parameter. Hence, instead of including deviations as a delayed second optimization objective, further research may involve using $d_{max}^+ = 0$ at the beginning and to gradually increase the value to admit more and more deviations as the optimization progresses.

6.2   Scenario 2: Roles, Deviations, Compliance Score, License Costs

In the second evaluation scenario, all previously presented objectives are included into the fitness function $f : \Pi_F \to \mathbb{R}^4$ of the addRole-EA, such that $f(\pi) = (|R|,\ d(RUPA, UPA),\ CS(\pi),\ LC(\pi))^T$, resulting in a four-dimensional optimization problem (4D-approach). Again, $\Pi_f = \Pi^+$ is considered. In a first step,again, the effect of delayed admittance of deviations is examined. For this

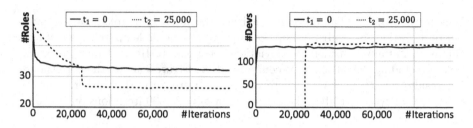

Fig. 7. Average number of roles/deviations considering t_1 resp. t_2 on PS_02.

purpose, 20 runs of the MO-addRole-EA were performed with different random seeds once permitting deviations immediately, i.e. as of iteration $t_1 = 0$ and once permitting deviations delayed, i.e. as of iteration $t_2 = 25,000$. Again, all non-dominated role concepts were selected from the individuals obtained from the execution of 100,000 iterations. Figure 8 shows the corresponding results for $d^+_{max} = 0.5$ on PS_02 and PS_05. Since it is no longer possible to represent role concepts in a two-dimensional Cartesian coordinate system, they are represented using parallel coordinates. Similar to Scenario 1, the delayed admittance of deviations leads to significantly better results in terms of the number of roles. In addition, the values obtained for the other optimization objectives have not worsened noticeably. In some cases, they have even improved.

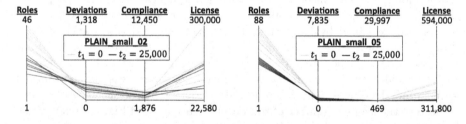

Fig. 8. Immediate (t_1) and delayed (t_2) admittance of deviations for 4D-approach.

Table 1 provides more insights showing the number of roles, the number of deviations as well as the compliance scores and the license costs at $t = 100,000$. For each of these optimization objectives, the minimum (min.), the maximum (max.), the average (avg.) and the standard deviation (SD) across all non-dominated role concepts obtained from the 20 runs of the MO-addRole-EA are given.

Section 4 dwelt on the relationship between compliance score, license costs and deviations. It was shown that positive deviations tend to increase the compliance score and the license costs. However, it is also possible that a positive deviation has no influence on the two optimization objectives. Such deviations are not critical neither from an economic nor from a safety perspective. Therefore, it is interesting to further explore another role mining scenario in which

Table 1. Comparing Immediate and Delayed Admittance of Deviations

		Roles		Deviations		Compliance Score		License Costs	
		t_1	t_2	t_1	t_2	t_1	t_2	t_1	t_2
PS_02	Avg.	32.91	22.00	166.91	175.71	2,430.18	2,505.43	253,036.36	245,200.00
	Min.	19	16	0	0	1,876	1,876	225,800	225,800
	Max.	44	27	308	262	3,119	2,955	297,200	260,800
	SD	8.04	3.46	91.22	80.92	372.78	340.64	19,576.02	12,620.62
PS_05	Avg.	74.45	46.00	169.09	105.00	500.64	478.00	340,436.36	313,200.00
	Min.	59	42	0	0	469	469	311,800	311,800
	Max.	84	50	283	263	566	512	379,400	316,000
	SD	7.34	2.58	84.70	88.60	28.48	13.51	21,742.09	1,746.11

positive deviations are admitted, but the number of deviations is not included as optimization objective. At this, it is expected that the inclusion of compliance score and license costs also regulates the number of deviations. This results in a three-dimensional optimization problem (3D-approach) with the following fitness function: $f : \Pi^+ \to \mathbb{R}^3$, $f(\pi) = (|R|, CS(\pi), LC(\pi))^T$.

In the following, the 4D- and the 3D-approach are evaluated and compared with each other. For this purpose, 20 repetitions with different random seeds of the MO-addRole-EA are executed with either the 3D- or 4D-approach. Figure 9 shows the non-dominated role concepts for $t_1 = 0$ and $d_{max}^+ = 0.5$ on PS_02 as well as PS_05 after 100,000 iterations considering all non-dominated role concepts. Apparently, removing the number of deviations as optimization objective has almost no effect on the values of the different optimization objectives on PS_02. On benchmark instance PS_05, however, removing the number of deviations as optimization objective significantly decreased the number of roles. Additionally, the compliance scores and the license costs of the resulting role concepts are reduced, whereas the associated number of deviations is increased.

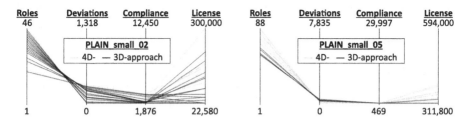

Fig. 9. Comparison of 3D- and 4D-approach.

This is reflected in Table 2, which shows the number of roles, the number of deviations as well as the values of the compliance score and the license costs at $t = 100,000$. The same experiments were also performed for delayed deviations

Table 2. Comparing 3D- and 4D-approach

		Roles		Deviations		Compliance Score		License Costs	
		t_1	t_2	t_1	t_2	t_1	t_2	t_1	t_2
PS_02	Avg.	32.91	34.24	166.91	155.82	2,430.18	2,175.06	253,036.36	238,647.06
	Min.	19	19	0	3	1,876	1,876	225,800	225,800
	Max.	44	43	308	284	3,119	3,061	297,200	274,800
	SD	8.04	6.27	91.22	91.49	372.78	379.58	19,576.02	15,034.32
PS_05	Avg.	74.45	58.40	169.09	296.20	500.64	470.60	340,436.36	314,600.00
	Min.	59	55	0	222	469	469	311,800	311,800
	Max.	84	63	283	392	566	473	379,400	325,800
	SD	7.34	3.07	84.70	57.26	28.48	1.96	21,742.09	5,600.00

which were admitted as of iteration $t_2 = 25,000$. In this case, the 4D-approach and the 3D-approach led to similar results considering the values of the different optimization objectives on PS_02 as well as on PS_05.

In conclusion, it can be stated that both, removing deviations as optimization objective as well as delaying the admittance of deviations have the potential to improve the results in some of the evaluation scenarios. Especially in industrial scenarios where further objectives may be involved, resulting in an even higher dimensional fitness function, the removal of deviations from the set of optimization objectives may be a sensible technique to reduce the dimensionality of the fitness function. If special focus is set on finding a minimal number of roles, it is advisable to first apply the Basic RMP for some initial iterations, hence to delay the admittance of deviations, before proceeding with MO role mining.

7   Conclusion and Future Works

In this paper, different aspects of MO role mining were investigated. First different potential optimization objectives were reviewed and a general definition of the *Multi-objective Role Mining Problem* was provided. Subsequently, an evolutionary algorithm, the addRole-EA, designed for the single-objective Basic Role Mining Problem was adapted to be applicable in MO role mining scenarios. Based on this, different aspects of MO role mining were investigated in different evaluation scenarios. It could be shown that in order to improve the role concepts obtained from the application of the MO-addRole-EA with respect to the number of roles, it is worthwhile to apply the Basic RMP for some initial iterations before including further optimization objectives. To examine the relation between the delay duration and the achievable solution qualities is subject to future works. Moreover, it could be shown that removing the number of deviations as optimization objective can contribute to improved results in some scenarios, which can be relevant in real-world scenarios where an even higher dimensional fitness function may be involved. Since this paper was primarily concerned with highlighting the different effects within MO role mining, one aspect for future considerations consists of the examination of performance aspects. First, the algorithm itself and with that the dependencies of the effects shown on the operators of the MO-addRole-EA must be examined. Thereafter, the MO-addRole-EA should be compared against other state-of-the-art MO optimization techniques, such as MOEA/D.

References

1. PwC, PwC's Global Economic Crime and Fraud Survey 2022. PricewaterhouseCoopers (2022)
2. Sandhu, R.S., Coyne, E.J., Feinstein, H.L., Youman, C.E.: Role-based access control models. Computer **29**(2), 38–47 (1996)
3. Vaidya, J., Atluri, V., Guo, Q.: The role mining problem. In: Proceedings of the 12th ACM SACMAT, Sophia Antipolis, France, 20–22 June 2007, pp. 175–184 (2007)
4. Blundo, C., Cimato, S.: A simple role mining algorithm. In: Proceedings of the ACM SAC, Sierre, Switzerland, 22–26 March 2010, pp. 1958–1962 (2010)
5. Kumar, R., Sural, S., Gupta, A.: Mining RBAC roles under cardinality constraint. In: Jha, S., Mathuria, A. (eds.) ICISS 2010. LNCS, vol. 6503, pp. 171–185. Springer, Heidelberg (2010). https://doi.org/10.1007/978-3-642-17714-9_13
6. Molloy, I.M., et al.: Evaluating role mining algorithms. In: Proceedings ACM SACMAT 2009, Stresa, Italia, 3–5 June 2009, pp. 95–104 (2009)
7. Schlegelmilch, J., Steffens, U.: Role mining with ORCA. In: Proceedings ACM SACMAT 2005, pp. 168–176. ACM Press, New York (2005)
8. Vaidya, J., Atluri, V., Warner, J., Guo, Q.: Role engineering via prioritized subset enumeration. IEEE Trans. Dependable Secure Comput. **7**(3), 300–314 (2010)
9. Huang, H., Shang, F., Liu, J., Du, H.: Handling least privilege problem and role mining in RBAC. J. Comb. Optim. **30**(1), 63–86 (2015)
10. Ene, A., Horne, W., Milosavljevic, N., Rao, P., Schreiber, R., Tarjan, R.E.: Fast exact and heuristic methods for role minimization problems. In: Proceedings of the ACM Symposium on Access Control Models and Technologies - SACMAT 2008, pp. 1–10. ACM Press, New York (2008)
11. Zhang, D., Ramamohanarao, K., Ebringer, T.: Role engineering using graph optimisation. In: Proceedings of the ACM Symposium on Access Control Models and Technologies - SACMAT 2007, pp. 139–144. ACM Press, New York (2007)
12. Zhang, D., Ramamohanarao, K., Versteeg, S., Zhang, R.: Graph based strategies to role engineering. In: Proceedings of the Sixth Annual Workshop on Cyber Security and Information Intelligence Research - CSIIRW 2010, pp. 1–4. ACM Press, New York (2010)
13. Mitra, B., Sural, S., Vaidya, J., Atluri, V.: A survey of role mining. ACM Comput. Surv. **48**(4), 1–37 (2016)
14. Saenko, I., Kotenko, I.: Genetic algorithms for role mining problem. In: Proceedings of the 19th PDP 2011, Ayia Napa, Cyprus, 9–11 February 2011, pp. 646–650 (2011)
15. Du, X., Chang, X.: Performance of AI algorithms for mining meaningful roles. In: Proceedings of the IEEE Congress on Evolutionary Computation, CEC 2014, Beijing, China, 6–11 July 2014, pp. 2070–2076 (2014)
16. Anderer, S., Kreppein, D., Scheuermann, B., Mostaghim, S.: The addRole-EA: a new evolutionary algorithm for the role mining problem. In: Proceedings of the 12th IJCCI 2020, Budapest, Hungary, 2–4 November 2020, pp. 155–166 (2020)

17. Anderer, S., Scheuermann, B., Mostaghim, S., Bauerle, P., Beil, M.: RMPlib: a library of benchmarks for the role mining problem. In: SACMAT 2021: Proceedings of the 26th ACM SACMAT, Virtual Event, Spain, 16–18 June 2021, pp. 3–13 (2021)

18. Anderer, S., Alpay, S., Scheuermann, B., Mostaghim, S.: On using authorization traces to support role mining with evolutionary algorithms. In: Proceedings of the 14th IJCCI 2022, Valletta, Malta, 24–26 October 2022, pp. 121–132 (2022)

19. Islam, A.K., et al.: Fraud detection in ERP systems using scenario matching. In: Security and Privacy - Silver Linings in the Cloud - Proceedings of the 25th IFIP TC-11 International Information Security Conference, SEC 2010, Held as Part of WCC 2010, Brisbane, Australia, 20–23 September 2010, vol. 330, pp. 112–123 (2010)

20. Saenko, I., Kotenko, I.: Using genetic algorithms for design and reconfiguration of RBAC schemes. In: Proceedings of the 1st International Workshop on AI for Privacy and Security, PrAISe@ECAI 2016, The Hague, Netherlands, 29–30 August 2016, pp. 1–9 (2016)

21. Colantonio, A., Di Pietro, R., Ocello, A.: A cost-driven approach to role engineering. In: Proceedings of the 2008 ACM Symposium on Applied Computing (SAC), Fortaleza, Ceara, Brazil, 16–20 March 2008, pp. 2129–2136 (2008)

22. Xu, Z., Stoller, S.D.: Algorithms for mining meaningful roles. In: Proceedings of the 17th ACM SACMAT, Newark, NJ, USA, 20–22 June 2012, pp. 57–66 (2012)

23. Molloy, I.M., et al.: Mining roles with semantic meanings. In: Proceedings of the 13th ACM SACMAT, Estes Park, CO, USA, 11–13 June 2008, pp. 21–30 (2008)

24. Molloy, I.M., et al.: Mining roles with multiple objectives. ACM Trans. Inf. Syst. Secur. (TISSEC) 1–35 (2010)

25. Stoller, S.D., Bui, T.: Mining hierarchical temporal roles with multiple metrics. J. Comput. Secur. 121–142 (2018)

26. Deb, K., Pratap, A., Agarwal, S.: A fast and elitist multiobjective genetic algorithm: NSGA-II. IEEE Trans. Evol. Comput. $6(2)$, 182–197 (2002)

Greedy Heuristic Guided
by Lexicographic Excellence

Satya Tamby$^{(\boxtimes)}$ (ID), Laurent Gourvès (ID), and Stefano Moretti (ID)

LAMSADE, CNRS, Université Paris-Dauphine, Université PSL, 75016 Paris, France
{satya.tamby,laurent.gourves,stefano.moretti}@lamsade.dauphine.fr

Abstract. This article deals with a basic greedy algorithm which, element by element, is able to construct a feasible solution to a wide family of combinatorial optimization problems. The novelty is to guide the greedy algorithm by considering the elements of the problem by order of merit, following a social ranking method. Social rankings come from social choice theory. The method used in the present article, called lexicographic excellence, sorts individual elements on the basis of the performances of groups of elements. In order to validate our approach, we conduct a theoretical analysis on matroid optimization problems, followed by a thorough experimental study on the multi-dimensional knapsack and the maximum weight independent set problem, leading to promising results.

Keywords: Combinatorial optimization · Greedy heuristic · Social ranking · Lexicographic excellence

1 Introduction

A greedy heuristic constructs a solution element by element, myopically choosing at each step the next element that provides the highest benefit. For instance, the elements can be initially sorted based on their individual benefits (e.g., Kruskal' algorithm [16]). However, in general, one could argue that the elements should be initially ordered in a different way, for example, according to their capacity to provide higher benefits when they interact with other elements. So, an alternative strategy is arranging the elements by taking into account the collective benefit generated by subsets (referred to as "coalitions") of elements. As for the strategy of sorting elements based on their individual benefits, the precise value of a coalition can be unimportant in building up a global optimal solution, and only the ordinal comparison of coalitional benefits matters. Moreover, elements may belong to several coalitions and a method to convert the information on coalitional benefits into an individual ranking is required to guide the greedy heuristic. To this purpose, our approach relies on the application of the notion

Supported by Agence Nationale de la Recherche (ANR), project THEMIS ANR-20-CE23-0018.

© The Author(s), under exclusive license to Springer Nature Switzerland AG 2024
T. Stützle and M. Wagner (Eds.): EvoCOP 2024, LNCS 14632, pp. 97–112, 2024.
https://doi.org/10.1007/978-3-031-57712-3_7

of *social rankings* [5, 12, 17], that has been recently introduced and studied in the field of social choice theory with the aim to rank single elements (or *individuals*) according to their position in a ranking over coalitions. More precisely, a social ranking is a map that assigns to each ranking over subsets of a finite set N another ranking over the single elements of N reflecting their overall influence to produce the ranking over the subsets.

Several social rankings have been proposed in the literature, and one of these is the *lexicographic excellence (lex-cel)* [5], which is based on the idea that the most influential individuals are those who occur most frequently in the highest positions in the ranking over coalitions. To be more specific, the lex-cel proceeds in a lexicographic way over the equivalence classes of a ranking over coalitions: the elements that occur the most in coalitions within the best equivalence class are placed at the top of the individual ranking; in case of a tie, the lex-cel examines the occurrences of elements in the coalitions of the second-best equivalence class, using the same principle to break the ties in favour of the elements that occur the most; and so on, till all ties are broken or the last equivalence class is reached (and the elements still on a tie at this point are declared indifferent).

Clearly, due to the exponential number of possible coalitions, the computational cost to produce a ranking over all coalitions is prohibitive in our setting, as it would require solving an optimization problem for the computation of the benefit of each coalition. For this reason, in this paper we introduce a generalized notion of lex-cel that applies to a restricted family of coalitions, possibly to those of sufficiently small size, where the computational effort to solve the corresponding optimization problem is not too demanding. At the same time, to better homogenise the coalitional effect of relatively close coalitions, we adopt alternative discretization approaches to generate larger equivalence classes formed by coalitions of "similar" value. In our framework, we believe that the reason why lex-cel turns out to be effective in guiding a greedy heuristic is twofold. First, the elements belonging to the best coalitions can likely prove to be more effective to take part of the global optimum than those based on a purely individual benefit. Second, a lexicographic approach is less sensitive to small fluctuations in the individual values, and therefore the ordering of elements for a greedy strategy is more robust to variations due to random generation of numerical instances.

In the first part of this paper, we show that for some specific combinatorial problems on weighted matroids, a greedy heuristic guided by the lex-cel computed over all possible coalitions of items may generate an optimal solution in general. However, even though these results provide a theoretical support to the ability of the lex-cel to properly guide a greedy heuristic, considering the exponential number of all possible coalitions, it seems challenging to use the exact lex-cel in practice. Therefore, in the second part of this paper we propose and compare multiple "approximation' strategies aimed at reducing the computational burden of our approach. Essentially, our strategy to reduce the computational burden works through three independent directions: 1) sampling a limited number of coalitions for the generation of sub-instances of the optimization problem; 2) reducing the number of equivalence classes in the coalitional ranking and, consequently, the number of lexicographic comparisons to compute the lex-cel; 3) exploiting the fact that, thanks to the monotonicity of the objective function, each feasible solution can guide a greedy algorithm to compute the optimal value of sub-instances without having to resort to more computationally expensive methods. In parallel, to improve the quality of the lex-cel approximation, we test different methods for sampling coalitions as well as alternative discretization criteria to reduce the number of equivalence classes. Our methodology has been experimentally tested over the generation of many instances of two well-known combinatorial optimization problems: *multi-dimensional knapsack* and *max-weight independent set*. We performed several experiments to evaluate our approach based on the lex-cel over multiple scenarios for those two combinatorial optimization problems. Our experimental analysis suggests that the use of the lex-cel computed on a quite limited number of coalitions of small size outperforms classical greedy heuristics from the literature [11,13].

The roadmap of the paper is as follows. After discussing some related articles from the literature in the next section, in Sect. 2 we introduce some preliminaries on our framework. Section 3 is devoted to the theoretical analysis of the performance of the exact lex-cel to guide a greedy algorithm on combinatorial optimization problems over weighted matroids. Section 4 deals with the explanation of our strategy to reduce the computational burden of the lex-cel via a proxy of the ranking. In Sect. 5 we provide an experimental analysis of our approach on the multi-dimensional knapsack problem and on the max-weight independent set problem. Section 6 concludes with some future directions.

Due to space constraints, some material is missing and will appear in an extended version.

1.1   Related Work

Social Ranking. Apart from the lex-cel, other lexicographic social rankings have been proposed in the related literature. In [5] a dual definition of the lex-cel has been axiomatically studied with the objective to punish mediocrity. Like the lex-cel, the *dual lex-cel* compares lexicographically the vector of occurrences in the equivalence classes in the ranking over coalitions but, different from the lex-cel, the dual one starts from the worst equivalence class and breaks ties punishing individuals more represented in the worst equivalence classes. In [1], the authors have introduced two extensions of the lex-cel aimed at rewarding the excellence of individuals and taking into account the size of coalitions in a way that small coalitions count less. In a similar way, in the paper [4], two other lexicographic social rankings are considered where the occurrence in coalitions is considered only in case singletons belong to the same equivalence classes in the ranking over coalitions. Moreover, there exist alternative notions of social ranking that do not involve any lexicographic comparison. For instance, the *CP-majority social ranking* [12] ranks two individuals i and j as one would do using a voting procedure where the voters are the coalitions containing either i or j. Another family of social rankings takes inspiration from well-known concepts from cooperative game theory like the Banzhaf value [2,14] and the core stability notion for hedonic games [3,10].

Greedy Approach in Combinatorial Optimization. Greedy algorithms are basic techniques for combinatorial optimization problems [9,20]. They can find an optimum in single matroid optimization problems (including the weighted spanning tree problem) [15,16] and also in various scheduling problems such as the minimization of the weighted sum of completion times of jobs to be executed on a single machine [7]. Greedy algorithms do not provide an optimal solution for every combinatorial optimization problem but they sometimes offer some performance guarantee; the value of its output is guaranteed to be within a non trivial multiplicative factor (a.k.a. approximation ratio) from the optimal value [22]. This is for example the case for well-known combinatorial problems such as *set cover* [9], *knapsack* [13], *independent set* [11] or *max-cut* [21].

Greedy algorithmic techniques also appear in involved (meta-)heuristics whose performance can be evaluated empirically [18].

2   Our Framework

2.1   A General Optimization Problem and a Greedy Algorithm

We consider a general combinatorial optimization problem P defined on a given ground set of elements $\mathcal{E} = \{e_1, \ldots, e_n\}$. A *feasible solution* is a subset of $\mathcal{E}$ satisfying some constraints. Let $\mathcal{F}$ denote the set of feasible solutions of P. We assume $\emptyset \neq \mathcal{F} \subset 2^{\mathcal{E}}$. Namely, a feasible solution is a subset of $\mathcal{E}$ and the problem P admits at least one feasible solution. A *partial solution* $\tilde{s}$ is a subset of $\mathcal{E}$ for which there exists $s \in \mathcal{F}$ satisfying $\tilde{s} \subseteq s$. Let $\tilde{\mathcal{F}}$ denote the set of partial solutions of P. Thus, $\emptyset \in \tilde{\mathcal{F}}$ and $\mathcal{F} \subset \tilde{\mathcal{F}}$. We are equipped with an objective function $f : 2^{\mathcal{E}} \to \mathbb{R}$ which associates a value to every subset of $\mathcal{E}$. The objective of the general combinatorial optimization problem P is to find $s^* \in \mathcal{F}$ so that $f(s^*)$ is maximized.

Many important problems in combinatorial optimization fall in the framework of P, such as the search of a maximum weight base of a *matroid*, (more details on matroids are given in Sect. 3), the (multi-dimensional) knapsack problem [13], or the max-weight independent set problem [11] (the last two problems are defined in Sect. 5). For these problems, the objective function is nondecreasing monotone, i.e., $f(s) \leq f(s')$ whenever $s \subseteq s'$.

We consider a simple greedy algorithm for constructing a feasible solution s to the general problem P (cf. [9,20] for a general introduction to greedy algorithms). The algorithm requires a ranking of the elements of $\mathcal{E}$. Let π be a permutation of $i \in [n] = \{1, \ldots, n\}$. Given any $i \in [n]$, $e_{\pi(i)}$ is the i-th element of $\mathcal{E}$ according to π. The greedy algorithm is described in Algorithm 1. Starting from scratch, a solution is gradually and myopically constructed by adding elements of $\mathcal{E}$ according to π's order. The insertion of an element is an irrevocable decision.

The time complexity of Algorithm 1 mostly depends on line 3 where one tests if a subset of $\mathcal{E}$ is a partial solution. It is not difficult to see that Algorithm 1 always builds a feasible solution for our general combinatorial optimization problem P.

Algorithm 1: Greedy algorithm

1 $s \leftarrow \emptyset$
2 **for** $i = 1$ **to** n **do**
3   **if** $s \cup \{e_{\pi(i)}\} \in \tilde{\mathcal{F}}$ **then**
4         $s \leftarrow s \cup \{e_{\pi(i)}\}$

5 **return** s

The value of the output of Algorithm 1 heavily depends on π which can be determined with the help of any information on P. For example, one can sort the elements of $\mathcal{E}$ in such a way that $\pi(i) \leq \pi(j) \Leftrightarrow f(\{e_{\pi(i)}\}) \geq f(\{e_{\pi(j)}\})$, as for Kruskal's algorithm. However, π is not necessarily fixed in advance;[1] at step i, $\pi(i)$ can be determined with the help of $\{\pi(1), \ldots, \pi(i-1)\}$ and the current (partial) solution (as in the greedy algorithm for the unweighted independent set problem which takes, at each round, the node v of minimum degree, and deletes v together with its neighbors [11]). An important feature of π is that it carries an *ordinal information* of the elements. Then, which ordinal information makes the greedy algorithm an efficient algorithm? Many articles deal with this question and most of them consider sorting the elements of $\mathcal{E}$ according to some profit-to-cost ratio (later called *efficiency* in Sect. 5).

We propose to follow a new approach which consists of using a social ranking method to guide the greedy algorithm for solving the general combinatorial problem P. More specifically, the focus of this article is on lexicographic excellence [5] (lex-cel in short) defined in Sect. 2.2. In a nutshell, lex-cel considers a ground set of elements, and ranks all the subsets of elements (a.k.a. coalitions) by strength with possible ties. Then, lex-cel outputs a ranking on the elements which depends on the ranking on the subsets of elements.

Going back to Algorithm 1 and problem P, we propose to use the lex-cel to rank the elements of $\mathcal{E}$ on the basis of the performances of the subsets of $\mathcal{E}$. Algorithm 1 is then equipped with an ordering π prescribed by the lex-cel. In order to compute the lex-cel ranking of $\mathcal{E}$, the performance of every subset of elements $X \in 2^{\mathcal{E}}$ is simply defined as $\max_{\{s \in \tilde{\mathcal{F}} | s \subseteq X\}} f(s)$. Here, each $X \in 2^{\mathcal{E}}$ induces a sub-instance of the original instance of P, where $\{s \in \tilde{\mathcal{F}} \mid s \subseteq X\}$ is the set of partial solutions of the sub-instance.

Unfortunately, it is computationally prohibitive to compute the exact ranking of the lex-cel, simply because it takes all possible subsets of $\mathcal{E}$ into consideration. In the following, we distinguish Algorithm 1 equipped with the *exact version* of

---

[1] π may change over time.

the lex-cel (where all the possible coalitions are taken into account, as well as all possible values taken by these coalitions) from its *relaxed* version (where only a sample of possible coalitions is taken into account, and coalitions of different but close values can be considered to have the same strength). With the exact version, the execution time of Algorithm 1 is prohibitive, but we are interested in the quality of the solutions that it outputs, in order to validate the approach of guiding a greedy with a social ranking method. With the relaxed version, the running time of Algorithm 1 is reasonable, and we aim at empirically compare the quality of its output with other standard greedy algorithms.

2.2   Lexicographic Excellence

The collection of all nonempty subsets (also known as *coalitions*) of $[n] = \{1, \ldots, n\}$ is represented by $\mathcal{P}([n])$. We use the symbol $\succsim_{\mathcal{C}}$ to refer to a *ranking* (also called *total preorder*) on the collection of sets $\mathcal{C} \subseteq \mathcal{P}([n])$, and it is called *coalitional ranking* on $\mathcal{C}$. For any pair of sets $S, T \in \mathcal{C}$, the notation $S \succsim_{\mathcal{C}} T$ will be understood as the statement "coalition S is at least as powerful as coalition T" considering the coalitional ranking $\succsim_{\mathcal{C}}$. The *symmetric* part of $\succsim_{\mathcal{C}}$ will be denoted by $\sim_{\mathcal{C}}$, indicating that for any S and T in $\mathcal{C}$, $S \sim_{\mathcal{C}} T$ if and only if $S \succsim_{\mathcal{C}} T$ and $T \succsim_{\mathcal{C}} S$. On the other hand, the *asymmetric* part of $\succsim_{\mathcal{C}}$ will be represented by $\succ_{\mathcal{C}}$, denoting that for any S and T in $\mathcal{C}$, $S \succ_{\mathcal{C}} T$ if and only if $S \succsim_{\mathcal{C}} T$ and it is not true that $T \succsim_{\mathcal{C}} S$. Therefore, considering any S and T in $\mathcal{C}$, $S \succ_{\mathcal{C}} T$ indicates that "coalition S is strictly more powerful than coalition T" while $S \sim_{\mathcal{C}} T$ means that "coalitions S and T are equally powerful".

Consider a coalitional ranking $\succsim_{\mathcal{C}}$ in the following form: $S_1 \succsim_{\mathcal{C}} S_2 \succsim_{\mathcal{C}} \ldots \succsim_{\mathcal{C}} S_{|\mathcal{C}|}$. The quotient order of $\succsim_{\mathcal{C}}$ is represented as $\Sigma_1 \succ_{\mathcal{C}} \Sigma_2 \succ_{\mathcal{C}} \ldots \succ_{\mathcal{C}} \Sigma_l$, where the subsets S_j are grouped into equivalence classes Σ_k generated by the symmetric part of $\succsim_{\mathcal{C}}$. Given a coalitional ranking $\succsim_{\mathcal{C}}$ and its associated quotient order $\Sigma_1 \succ_{\mathcal{C}} \Sigma_2 \succ_{\mathcal{C}} \ldots \succ_{\mathcal{C}} \Sigma_l$, we denote as i_k the number of sets in Σ_k that contain the element i, where $k = 1, \ldots, l$. Additionally, $\theta_{\mathcal{C}}^{\succsim}(i)$ (or, simply, $\theta(i)$, if $\succsim_{\mathcal{C}}$ is clear from the context) refers to the l-dimensional vector $\theta^{\succsim_{\mathcal{C}}}(i) = (i_1, \ldots, i_l)$ associated with $\succsim_{\mathcal{C}}$. The lexicographic order $\geq_L$ among vectors $\theta^{\succsim_{\mathcal{C}}}(i) = (i_1, \ldots, i_l)$ and $\theta^{\succsim_{\mathcal{C}}}(j) = (j_1, \ldots, j_l)$ is defined as follows: $\theta^{\succsim_{\mathcal{C}}}(i) \geq_L \theta^{\succsim_{\mathcal{C}}}(j)$ if either $\theta^{\succsim_{\mathcal{C}}}(i) = \theta^{\succsim_{\mathcal{C}}}(j)$ or there exists t such that $i_t > j_t$ and $i_r = j_r$ for all $r \in \{1, \ldots, t-1\}$.

Definition 1 ((Generalized) lexicographic-excellence (lex-cel)).
For any coalitional ranking $\succsim_{\mathcal{C}}$ with $\mathcal{C} \subseteq \mathcal{P}([n])$, the *generalized lexicographic-excellence (lex-cel)* is the ranking $R_{gle}^{\succsim_{\mathcal{C}}}$ on $[n]$ such that:

$$iR_{gle}^{\succsim_{\mathcal{C}}} j \iff \theta^{\succsim_{\mathcal{C}}}(i) \geq_L \theta^{\succsim_{\mathcal{C}}}(j).$$

for all $i, j \in [n]$ (in the following, $I_{gle}^{\succsim_{\mathcal{C}}}$ and $P_{gle}^{\succsim_{\mathcal{C}}}$ stand for the symmetric part and the asymmetric part of $R_{gle}^{\succsim_{\mathcal{C}}}$, respectively).

Notice that Definition 1 generalizes the notion of lex-cel introduced in [5] in the sense that it applies to chains over any collection of sets in $\mathcal{P}([n])$ (and not necessarily to the maximal chain $\succsim_{\mathcal{P}([n])}$ over all nonempty subsets in $\mathcal{P}([n])$, as done in [5]).

Example 1. Let $[n] = \{1, 2, 3\}$ and consider the coalitional ranking $\succsim_{\mathcal{C}}$ with $\mathcal{C} = \{\{1, 3\}, \{1, 2\}, \{2, 3\}, \{1\}\}$ and such that: $\{1, 3\} \succ_{\mathcal{C}} \{1, 2\} \sim_{\mathcal{C}} \{2, 3\} \succ_{\mathcal{C}} \{1\}$. The quotient order of $\succsim$ is $\Sigma_1 = \{\{1, 3\}\} \succ_{\mathcal{C}} \Sigma_2 = \{\{1, 2\}, \{2, 3\}\} \succ_{\mathcal{C}} \Sigma_3 = \{\{1\}\}$. We have $\boldsymbol{\theta}(1) = (1, 1, 1)$, $\boldsymbol{\theta}(2) = (0, 2, 0)$ and $\boldsymbol{\theta}(3) = (1, 1, 0)$. Notice that $\boldsymbol{\theta}(1) >_L \boldsymbol{\theta}(3) >_L \boldsymbol{\theta}(2)$. So, the lex-cel returns the following ranking: 1 $\boldsymbol{P}_{gle}^{\succsim_{\mathcal{C}}}$ 3 $\boldsymbol{P}_{gle}^{\succsim_{\mathcal{C}}}$ 2.

3   Greedy Guided by the Exact Version of Lex-cel

In this section, we study the performance of Algorithm 1 guided by the *exact* version of lex-cel (i.e., when the whole set of coalitions in $[n]$ are considered for the coalitional ranking; so, using the notation introduced in the previous section, $\mathcal{C} = \mathcal{P}([n])$) for some matroid optimization problems. Namely, we suppose that π in Algorithm 1 is equal to the lex-cel computed on $\succsim_{\mathcal{P}(\mathcal{E})}$. In this section, we do not consider the time complexity for eliciting π. The purpose is to justify that greedy with lex-cel is a promising approach in theory.

Before presenting our results, let us briefly define matroids optimization problems (see for example [15, 19] for more details) which fit in the general combinatorial optimization problem P described in Sect. 2.1.

In a matroid $(E, \mathcal{I})$, E is a set of elements and $\mathcal{I}$ is a collection of subsets of E called *independent sets*. $\mathcal{I}$ is such that (*i*) $\emptyset \in \mathcal{I}$, (*ii*) if $I_1 \in \mathcal{I}$, then $I_0 \subseteq I_1 \Rightarrow I_0 \in \mathcal{I}$, and (*iii*) if $(I_0, I_1) \in \mathcal{I}^2$, then $|I_0| < |I_1| \Rightarrow \exists e \in I_1 \setminus I_0$ such that $I_0 \cup \{e\} \in \mathcal{I}$. A *base* B of $(E, \mathcal{I})$ is an independent set of maximal cardinality, i.e., $B \in \mathcal{I}$ and $B \cup \{e\} \notin \mathcal{I}$ for all $e \in E \setminus B$. Given a weight function $w : E \rightarrow \mathbb{R}_{\geq 0}$, the weight of any $X \subseteq E$ is defined as $\sum_{e \in X} w(e)$. One can construct a base of $(E, \mathcal{I})$ of maximum weight with Algorithm 1, given that the elements of E are taken by non-increasing order of weight.[2] A base is said to be *optimal* if its weight is maximum over $\mathcal{I}$. A well-known matroid is the *graphic matroid*. The graphic matroid $(E_G, \mathcal{I}_G)$ associated with a connected and undirected graph $G = (V, A)$ is such that $E = A$, and $I \subseteq A$ belongs to $\mathcal{I}_G$ iff I is acyclic. A base of $(E_G, \mathcal{I}_G)$ is a spanning tree of G. An optimal base of $(E_G, \mathcal{I}_G)$ is a maximum weight spanning tree. Hence, the problem of finding an optimal base in a graphic matroid is nothing but the problem of finding a maximum weight spanning tree. Moreover, the greedy algorithm for computing an optimal base simply generalizes Kruskal's algorithm [16].

Because Algorithm 1 is optimal for matroid optimization problems when the elements are sorted by non-increasing weight, we propose to analyze the performance of Algorithm 1 equipped with the exact version of lex-cel on some matroid

---

[2] The maximization version of the matroid problem is considered but the minimization version is totally equivalent.

optimization problems. Our results are that it is optimal for some matroids but not for all of them.

A *cactus* is a connected graph in which any two simple cycles have at most one vertex in common.

Proposition 1. *Algorithm 1 equipped with the exact version of the lex-cel solves optimally instances of the maximum weight spanning tree such that the underlying graph is a cactus.*

Proof. In a cactus $G = (V, E)$, every edge belongs to at most one cycle. Thus, a maximum weight spanning tree of G can be built as follows: start from E and remove from each cycle exactly one of its lightest edges.[3] Since two distinct cycles of a cactus never share an edge, the decision of which edge is not retained in each cycle can be made independently.

If an edge e does not belong to any cycle, or if e belongs to a cycle C but its weight is not the smallest within C, then e must appear in every possible maximum weight spanning tree of G. Let us call them *heavy edges*. Edges that are not heavy are said to be *light*. If there are exactly q_C light edges in a cycle C of G, then each of them appears in $\frac{q_C - 1}{q_C}\tau$ maximum weight spanning trees where τ denotes the total number of maximum weight spanning trees in G. The first coordinate of the lex-cel vector of every light edge of C is $\frac{q_C - 1}{q_C}\tau$. The first coordinate of the lex-cel vector of every heavy edge is τ.

Therefore, Algorithm 1 equipped with the exact version of the lex-cel first picks the heavy edges, and they constitute a forest of G. Afterwards, the greedy algorithm considers the light edges, inserting them in the partial solution as long as no cycle is created. From the above discussion on the structure of a maximum weight spanning tree, we can conclude that the algorithm always outputs a maximum weight spanning tree since it excludes a light edge of every cycle.   □

Proposition 2. *There exists an edge weighted graph for which Algorithm 1 equipped with the exact version of lex-cel fails to output a spanning tree of maximum weight.*

Proof. Consider the instance depicted on Fig. 1. A tree of maximum weight has weight 10. Every coalition containing an optimal tree contains (B, D), plus (B, C) or (C, D) or $\{(B, C), (C, D)\}$, plus (A, B) or (A, C) or $\{(A, B), (A, C)\}$. Therefore, both (B, C) and (C, D) appear in 6 of the 9 coalitions of value 10. Similarly, both (A, B) and (A, C) appear in 6 of the 9 coalitions of value 10. The edge (B, D) appears in every coalition of value 10. The second best value for a coalition is 9. The only coalition having this value is $\{(A, B), (A, C), (B, D)\}$.

Therefore, we have $\boldsymbol{\theta}((B, D)) = (9, 1, \ldots)$, $\boldsymbol{\theta}((A, B)) = \boldsymbol{\theta}((A, C)) = (6, 1, \ldots)$, and $\boldsymbol{\theta}((B, C)) = \boldsymbol{\theta}((C, D)) = (6, 0, \ldots)$ which give

$$(B, D) \; \boldsymbol{P}_{gle}^{\succsim c}(A, B) \; \boldsymbol{I}_{gle}^{\sim c} \; (A, C) \; \boldsymbol{P}_{gle}^{\succsim c} \; (B, C)\boldsymbol{I}_{gle}^{\sim c} \; (C, D).$$

---

[3] The lightest edge of a cycle is not necessarily unique.

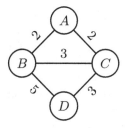

Fig. 1. A counter-example to Algorithm 1 with the exact version of the lex-cel.

The output of Algorithm 1 with this order on the edges is $\{(B, D), (A, B), (A, C)\}$ which is not optimal because its weight is 9.   □

Remark 1. Consider only coalitions of size at most two in the coalitional ranking generated on the graph depicted in Fig. 1, i.e. $\mathcal{C} = \{S \in \mathcal{P}([n]) : |S| \leq 2\}$. Note that both (B, C) and (C, D) appear in just one of the two best coalitions in $\mathcal{C}$ of value 8, while both (A, B) and (A, C) appear in just one of the two second-best coalitions in $\mathcal{C}$ of value 7. Therefore, we have $\theta^{\succsim^c}((B, D)) = (2, 2, \ldots)$, $\theta^{\succsim^c}((A, B)) = \theta^{\succsim^c}((A, C)) = (0, 2, \ldots)$, and $\theta^{\succsim^c}((B, C)) = \theta^{\succsim^c}((C, D)) = (2, 0, \ldots)$ which gives $(B, D) \, \mathbf{P}_{gle}^{\succsim^c} (B, C) \mathbf{I}_{gle}^{\succsim^c} (C, D) \, \mathbf{P}_{gle}^{\succsim^c} (A, B) \, \mathbf{I}_{gle}^{\succsim^c} (A, C)$. The output of Algorithm 1 with this order on the edges is $\{(B, D), (B, C), (A, B)\}$ which is an optimal tree. The strategy of limiting the lex-cel to small-sized coalitions will prove effective in general both from the point of view of the computational complexity of the method and from that of the quality of the generated solution over various optimization problems, as shown in Sect. 5.

Now let us consider the problem of finding a base of maximum weight in a general (i.e., non necessarily graphic) matroid. Our positive result about the greedy algorithm equipped with the exact version of the lex-cel is that it optimally solves instances admitting a small number of optimal bases. The omitted proof relies on Brualdi's matroid exchange property [6].

Proposition 3. *Algorithm 1 equipped with the exact version of the lex-cel solves optimally every instance of the maximum weight base of a matroid admitting at most three optimal bases.*

Proposition 3 applies, for example, to instances where the elements of the matroid have pairwise distinct weights, because the matroid admits a unique optimal base in this case. Proposition 3 cannot be extended to instances admitting at most four optimal bases because the instance depicted in Fig. 1 has exactly four optimal bases.

4   Greedy Guided by a Relaxed Version of the Lex-cel

In practice, the main obstacle for using the exact version of the lex-cel is its prohibitive running time. Therefore, we are going to describe a relaxed version of the lex-cel which relies on the following modifications: (i) consider only a subset of coalitions $\mathcal{C} \subset \mathcal{P}(\mathcal{E})$, and impose an upper bound on the cardinality of each coalition S in $\mathcal{C}$, (ii) target coalitions for which we can quickly know the optimal value, (iii) limit the number of equivalence classes of the lex-cel by grouping coalitions of close, but not necessarily equal, optimal value, and (iv) introduce a bias in the choice of the sub-collection coalitions $\mathcal{C}$.

We expect these modifications to significantly reduce the running time of the lex-cel, without deteriorating its capacity to guide the greedy algorithm towards efficient solutions.

Let us give more details about the aforementioned modifications. In general, it is not realistic to enumerate the whole set of sub-instances corresponding to all coalitions in $\mathcal{P}(\mathcal{E})$ and to evaluate them, because the number of sub-instances is exponential in $|\mathcal{E}|$. Consequently, our relaxed approach consists of considering a limited number of sub-instances. Moreover, since the computational cost of solving a sub-instance usually increases with its size, we propose to upper bound the size of the sub-instances. This means limiting the cardinality of the coalitions in $\mathcal{P}(\mathcal{E})$ that are considered in the generalized lex-cel. The pre-defined size of a coalition is referred to as its *depth*.

Concretely, to generate a sub-instance, one can sample a coalition S of a pre-defined cardinality from the ground set $\mathcal{E}$, and determine its value by computing the optimal solution of the problem P restricted to S (which is a partial solution of the global instance). The optimal solution of any sub-instance S can be found with a solver such as CPLEX. However, the described procedure is time consuming and we decided to resort to the following, more direct, method. It consists of sampling a permutation $\tilde{\pi}$ on a subset of elements $\tilde{S}$ of $\mathcal{E}$ whose cardinality does not exceed the depth. Afterwards, Algorithm 1 is used with $\tilde{\pi}$ to produce a partial solution S of the problem, where $S \subseteq \tilde{S}$. Then, provided that the objective function f of the general problem P is non-decreasing monotone ($f(A) \leq f(B)$ holds for all A, B such that $A \subseteq B \subseteq \mathcal{E}$), it is not difficult to see that the optimal value of S, $f(S)$, is reached by selecting S in its entirety. Since we will do experiments on problems with elements having non-negative values and weights (multidimensional knapsack, and max-weight independent set), they directly satisfy the monotonicity property.

Therefore, points (i) and (ii) are put into practice by executing Algorithm 1 with some sampled orderings on subsets of $\mathcal{E}$ whose cardinality does not exceed a prescribed depth.

Regarding point (iii), following the lex-cel consists of ordering the sampled coalitions of $\mathcal{C} \subset \mathcal{P}(\mathcal{E})$ according to their optimal value in order to generate a coalitional ranking $\succcurlyeq_{\mathcal{C}}$.

A second observation concerns the output of the lex-cel. Indeed, the number of equivalence classes corresponds to the number of different optimal values reached by the coalitions. Therefore, we may observe that if all optimal values are different, this algorithm amounts to chose greedily the items according to the rank of the coalitions they belong to. In other words, the items of the first coalition are taken and the solution is completed using the items of the second coalition and so on, which may be problematic since the last classes do not influence the final ranking. For this reason, we want to introduce some ties and thus limit the number of classes. This also has the advantage of controlling (i.e., upper bounding) the number of coordinates of the vector θ described in Sect. 2.2. Concretely, we suggest to regroup the coalitions that have close ranks. For this purpose, we split the coalitions into classes of (almost) the same cardinality, partitioning the coalitions into quantiles.

Finally, we explore the possibility of introducing a bias in the sampling of the coalitions. Indeed, we want to sample coalitions (corresponding to sub-instances) of high quality for the generalized lex-cel. For the determination of those coalitions, we sample a number of individuals equal to the depth, according to a distribution that is either uniform (all elements are picked with the same probability) or biased. In the biased distribution, the probability of selecting an individual depends on an *efficiency* measure already used in standard greedy algorithms (more details on the efficiency are provided in the following section).

5 Experiments

5.1 Problem Definitions and Instance Generation

Multi-dimensional Knapsack. We are given a set of n items. Each item $j \in \{1, \ldots, n\}$ has a positive value v_j and m non-negative weights $w_{ij}, i \in \{1, \ldots, m\}$. A maximal capacity $W_i \in \mathbb{R}_{>0}$ is provided for every dimension $i \in \{1, \ldots, m\}$. The goal of the *multi-dimensional knapsack* problem is to select a subset of items having the maximum total value without exceeding any of the capacities [8,13].

Several greedy algorithms have been proposed (cf. [13, Section 9.5.1]). An efficient way to sort the items is to estimate their *efficiency* $\text{eff}(j) = \frac{v_j}{\sum_{i=1}^{m} w_{ij}}$. A classical greedy algorithm, against which we are going to compare our heuristic guided by a proxy of the lex-cel, consists of applying Algorithm 1 where the items are taken by non-increasing order of efficiency (break ties arbitrarily) [13]. In the conducted experiments, values and weights are sampled uniformly in the set $\{1, \ldots, 50\}$ while we have $\max_{j \in \{1, \ldots, n\}} w_{ij} \leq W_i \leq \sum_{j=1}^{n} w_{ij}$ for each i in order to ensure that at least one item can be taken.

Max-weight Independent Set. Consider an undirected graph $G = (V, E)$ with $n = |V|$ and non-negative weights on its vertices. The weight of any vertex $i \in V$ is denoted by w_i. A subset of pairwise non-adjacent vertices is said to be *independent*. The goal of the max-weight independent set problem is to identify an independent set of vertices of G whose total weight is maximized.

A standard greedy algorithm (see for example [11]), against which we are going to compare our heuristic guided by a proxy of the lex-cel, iteratively considers the candidates (i.e., vertices that are not adjacent to any vertex that has already been taken), and computes their efficiency using the formula $\text{eff}(i) = \frac{w_i}{d(i)}$ where $d(i)$ denotes the degree of vertex i. Then, the vertex maximizing the efficiency is taken, and its neighbors are removed from the set of candidates. The algorithm pursues until no candidate is available. Note that the degree of the candidates must be re-computed at each iteration since some edges are removed each time a vertex is taken. In the conducted experiments, weights are sampled in $\{1, \ldots, 50\}$.

5.2   Analysis

Our approach has been implemented using the Haskell programming language[4] The experiments have been conducted on a NixOS virtual machine with 16 threads and $32GB$ of memory (note that our approach is not multi-threaded). The underlying processor is an AMD EPYC 7702 64-Core Processor.

In this section we synthesise our experiments over the *multi-dimensional knapsack* (Table 1) and *max weight independent set* problems (Table 2). We compare the performance of the greedy algorithm when it is guided by the generalized lex-cel under two distinct strategies for sampling coalitions (a *uniform* and a *biased* one) with the standard heuristic referred to as *standard*. In order to show that the lex-cel approaches do not amount to taking the best of the samples, comparisons also involve the maximum score reached by the coalitions using the biased and the uniform sampling, respectively referred to as *lexmax* and *biasedmax* in the tables.

We measure here the influence of three parameters: the number of quantiles, i.e., the size of the vector $\boldsymbol{\theta}$ (referred to as *#classes*), the number of samples that are generated (*#samples*) and the size of the samples (*depth*). Concerning the max weight independent set problem, the density of the graph also varies. Finally, tests are performed over instances of two different sizes. For each test, the optimal value y^* of the instance is computed. Then, given the value $\bar{y}$ provided by the selected approach, the optimality gap is computed using the formula: $\frac{100\bar{y}}{y^*}$. Each line shows the average optimality gap over 20 independent tests.

We first observe that the biased lex-cel improves significantly the standard greedy heuristic. This difference can be seen in particular on Table 2. This is also true for the uniform lex-cel, except for the knapsack of 200 items, with 50 classes, 1000 samples and a depth of 3. Our intuition is that, when the depth is too small, it is difficult to capture the interaction of the decisions.

The difference between the biased lex-cel and the uniform lex-cel is slight. It seems however that *biased* outperforms *uniform* on every test except in a limited number of cases, where the values are extremely close. In both cases, we note that the influence of the number of classes is not significant. The number of samples seems to induce a slight change, as we can expect that increasing the number of samples may improve the quality of the algorithm, but it is also not significant. However, samples of large depth provide coalitions of high quality which in turn improves the quality of the solution yielded by the lex-cel based approaches.

A last observation concerns the difference between *uniform* (resp., *biased*) and *lexmax* (resp., *biasedmax*). The output of the greedy algorithm guided by

---

[4] Our implementation is available at https://github.com/tambysatya/socialranking.

the lex-cel outperforms the best sampled coalition, which shows the capacity of our method to construct good solutions. However, the gap tends to reduce when the depth increases because in these situations, (nearly) inclusionwise maximal solutions are more likely enumerated. In particular, this can be observed when computing an independent set on dense graphs, where the inclusionwise maximal solutions typically have a very small cardinality compared to the size of the sampled solutions. Moreover, when the number of samples also grows, the overall quality of the solutions is improved. Hence, taking the best solutions in these settings leads to better results. However, it is important to notice that although it is computationally cheap to apply a greedy algorithm to score a single coalition, this quickly becomes time consuming when the number of coalitions grows. Therefore, we recommend sampling a reduced subset of small coalitions before aggregating their results using the lex-cel and constructing a single maximal solution.

Table 1. Knapsack problem with 200 items and 50 constraints.

#classes	#samples	depth	mean			lexmax	biasedmax
			uniform	biased	standard		
10	500	3	57.39	64.44	55.95	21.14	21.94
		5	65.15	74.21	60.02	35.84	39.74
		10	74.59	79.90	60.74	49.34	57.77
	1000	3	63.52	64.05	60.49	26.87	27.37
		5	67.30	76.07	61.56	36.23	39.57
		10	79.18	84.24	62.03	48.65	55.87
50	500	3	61.54	64.99	60.59	26.95	28.05
		5	63.38	65.55	60.08	34.87	38.18
		10	68.48	78.20	66.55	47.12	54.88
	1000	3	59.09	61.64	61.23	24.79	25.74
		5	69.94	74.01	62.50	42.99	46.91
		10	73.82	79.45	65.93	50.47	55.65

Table 2. Max-Weight independent set with 200 vertices.

density	#classes	#samples	depth	mean uniform	biased	standard	lexmax	biasedmax
0.1	10	500	3	79.61	84.23	71.89	10.86	11.39
			5	78.50	82.94	70.79	16.06	17.57
			10	79.64	86.23	69.11	24.97	29.90
		1000	3	80.10	85.98	68.31	10.82	11.25
			5	81.34	87.05	70.13	16.65	17.91
			10	82.80	87.21	71.22	25.74	31.34
	50	500	3	80.37	84.23	70.68	10.80	11.33
			5	78.91	82.39	69.13	16.19	17.97
			10	74.06	83.14	70.87	25.77	30.19
		1000	3	81.69	84.19	70.68	10.99	11.29
			5	78.33	85.51	70.62	16.92	18.14
			10	78.55	85.23	69.72	26.52	31.16
0.5	10	500	3	70.16	76.01	50.91	33.69	36.31
			5	78.36	79.31	51.99	41.56	47.99
			10	76.24	77.77	53.21	50.32	57.96
		1000	3	76.95	76.30	49.90	35.26	37.01
			5	74.67	79.24	52.13	42.13	47.96
			10	75.85	82.40	51.80	54.55	60.05
	50	500	3	72.64	74.98	46.98	34.88	36.81
			5	71.80	75.59	51.01	41.33	46.22
			10	72.87	78.37	53.14	50.41	58.29
		1000	3	72.96	75.17	50.70	35.52	37.13
			5	70.35	74.49	51.78	45.65	48.20
			10	75.94	79.92	51.61	53.00	58.51
0.7	10	500	3	72.59	78.10	52.41	45.19	51.44
			5	70.69	76.88	50.19	51.26	56.89
			10	71.92	81.06	51.38	59.01	68.42
		1000	3	74.09	75.50	52.86	48.33	52.45
			5	76.55	76.79	55.55	55.34	57.93
			10	72.85	76.55	48.22	60.39	69.55
	50	500	3	72.37	73.45	54.37	43.89	51.61
			5	69.76	79.50	54.86	52.38	58.45
			10	72.30	73.89	52.10	60.41	67.89
		1000	3	74.41	74.64	52.27	47.92	52.76
			5	71.02	78.12	52.16	54.64	60.14
			10	75.28	78.64	51.43	60.81	70.48

6   Conclusion and Future Work

The approach proposed in this article uses a social ranking method, called lex-cel, to guide a greedy algorithm in order to generate good feasible solutions to a general combinatorial problem. By restricting the computation of the generalized lex-cel to a sub-collection of partial solutions (i.e., the coalitions), the algorithm becomes tractable and our experiments indicate that it can outperform standard greedy methods. Finally, we also show that biasing the generation of the partial solutions can improve the quality of the output.

Further works may concern the study of a simulator that would generate good partial solutions, that can be aggregated using the generalized lex-cel in order to guide the greedy towards high-quality outputs. A possibility for the future is to use a neural network for the determination of the parameters of such a simulator which may be dependent on the instance.

References

1. Algaba, E., Moretti, S., Rémila, E., Solal, P.: Lexicographic solutions for coalitional rankings. Soc. Choice Welfare **57**(4), 817–849 (2021). https://doi.org/10.1007/s00355-021-01340-z
2. Banzhaf, J.F., III.: Weighted voting doesn't work: a mathematical analysis. Rutgers L. Rev. **19**, 317 (1964)
3. Béal, S., Ferrières, S., Solal, P.: A core-partition ranking solution to coalitional ranking problems. Group Decis. Negot. **32**, 1–21 (2023)
4. Béal, S., Rémila, E., Solal, P.: Lexicographic solutions for coalitional rankings based on individual and collective performances. J. Math. Econ. **102**, 102738 (2022)
5. Bernardi, G., Lucchetti, R., Moretti, S.: Ranking objects from a preference relation over their subsets. Soc. Choice Welfare **52**(4), 589–606 (2019). https://doi.org/10.1007/s00355-018-1161-1
6. Brualdi, R.: Comments on bases in dependence structures. Bull. Aust. Math. Soc. **1**(2), 161–167 (1969)
7. Brucker, P.: Scheduling Algorithms, 5th edn. Springer, Cham (2007)
8. Cacchiani, V., Iori, M., Locatelli, A., Martello, S.: Knapsack problems-an overview of recent advances. part ii: multiple, multidimensional, and quadratic knapsack problems. Comput. Oper. Res. **143**, 105693 (2022)
9. Cormen, T.H., Leiserson, C.E., Rivest, R.L., Stein, C.: Introduction to Algorithms. The MIT Press, Cambridge (2009)
10. Dreze, J.H., Greenberg, J.: Hedonic coalitions: optimality and stability. Econometrica: J. Econometric Soc., 987–1003 (1980)
11. Halldórsson, M.M., Radhakrishnan, J.: Greed is good: approximating independent sets in sparse and bounded-degree graphs. Algorithmica **18**(1), 145–163 (1997)
12. Haret, A., Khani, H., Moretti, S., Ozturk, M.: Ceteris paribus majority for social ranking. In: 27th International Joint Conference on Artificial Intelligence (IJCAI-ECAI-18), Stockholm, Sweden, pp. 303–309 (2018).https://doi.org/10.24963/ijcai.2018/42, https://hal.archives-ouvertes.fr/hal-02103421
13. Kellerer, H., Pferschy, U., Pisinger, D.: Knapsack Problems. Springer, Berlin (2004)
14. Khani, H., Moretti, S., Ozturk, M.: An ordinal Banzhaf index for social ranking. In: 28th International Joint Conference on Artificial Intelligence (IJCAI 2019), Macao, China, pp. 378–384 (2019).https://doi.org/10.24963/ijcai.2019/54, https://hal.archives-ouvertes.fr/hal-02302304
15. Korte, B., Vygen, J.: Combinatorial Optimization: Theory and Algorithms, 5th edn. Springer, Cham (2012)
16. Kruskal, J.B.: On the shortest spanning subtree of a graph and the traveling salesman problem. Proc. Am. Math. Soc. **7**(1), 48–50 (1956)

17. Moretti, S., Öztürk, M.: Some axiomatic and algorithmic perspectives on the social ranking problem. In: Rothe, J. (eds.) Algorithmic Decision Theory. Lecture Notes in Computer Science(), vol. 10576, pp. 166–181. Springer, Cham (2017). https://doi.org/10.1007/978-3-319-67504-6_12, https://hal.archives-ouvertes.fr/hal-02103398
18. Osman, I.H., Kelly, J.P. (eds.): Meta-Heuristics: Theory and Applications. Springer, New York (1996)
19. Oxley, J.: Matroid Theory. Oxford University Press, Oxford (2011)
20. Roughgarden, T.: Algorithms Illuminated (Part 3): Greedy Algorithms and Dynamic Programming. Soundlikeyourself Publishing, LLC, New York, NY (2019)
21. Sahni, S., Gonzalez, T.: P-complete approximation problems. J. ACM (JACM) **23**(3), 555–565 (1976)
22. Vazirani, V.V.: Approximation Algorithms. Springer, Cham (2001)

Reduction-Based MAX-3SAT with Low Nonlinearity and Lattices Under Recombination

Darrell Whitley[1]($\boxtimes$)(iD), Gabriela Ochoa[2](iD), Noah Floyd[1](iD),
and Francisco Chicano[3](iD)

[1] Department of Computer Science, Colorado State University, Fort Collins, USA
{whitley,noah.floyd}@colostate.edu
[2] Computing Science and Mathematics, University of Stirling, Stirling, Scotland
gabriela.ochoa@cs.stir.uk
[3] ITIS Software, University of Malaga, Malaga, Spain
chicano@uma.es

Abstract. A new construction is introduced for creating random MAX-3SAT instances with low nonlinearity. Instead of generating random clauses, we generate random SAT expressions over 3 variables and then convert these into CNF SAT clauses. We prove that this yields structured problems with much lower nonlinearity. We also introduce a new method for weighting MAX-SAT clauses that preserves low nonlinearity and also breaks up plateaus. We evaluate these new problems by enumeration of instances with $n = 30$ variables. One unexpected result is that Partition Crossover creates more tunnels on these semi-structured MAX-SAT problems compared to results on random NK landscapes. We show that Partition Crossover induces hypercube lattices over subsets of local optima; all of the local optima which appear in a lattice can be evaluated with a single linear equation.

Keywords: MAX-SAT · Recombination Operators · Partition Crossover · Evolutionary Combinatorial Optimization

1 Introduction

MAX-SAT optimization has a long history in the combinatorial optimization and AI communities. MAX-SAT solvers have applications in hardware and software verification, as well as in related problems of model checking, constraint satisfaction, subset selection, scheduling and planning problems [8,20].

In general, there are two broad classes of MAX-SAT solvers: exact methods and inexact heuristic methods. Modern exact solvers display excellent results in many real-world applications. Inexact solvers are typically based on stochastic local search algorithms and have their roots in earlier methods such as GSAT [25] and WalkSAT [26]; they display excellent results on random instances [3,24].

Typically, industrial satisfiability applications display lower nonlinearity compared to randomly generated MAX-SAT instances. This nonlinearity can be

© The Author(s), under exclusive license to Springer Nature Switzerland AG 2024
T. Stützle and M. Wagner (Eds.): EvoCOP 2024, LNCS 14632, pp. 113–128, 2024.
https://doi.org/10.1007/978-3-031-57712-3_8

measured by taking the Fourier transform of the problem instance and counting the number of nonlinear coefficients that appear in the resulting polynomial.

This paper has two goals. First, deterministic forms of recombination have been developed in the field of evolutionary computation. What can we transfer from work on the Traveling Salesman Problem and on k-bounded pseudo-Boolean functions to the MAX-SAT domain? We also discuss new opportunities for improving evolutionary algorithms for MAX-SAT.

Second, how can we generate random MAX-SAT problems that are more like real-world problems? This second goal is also motivated by renewed interest in transforms that can convert highly nonlinear Boolean functions into a quadratic form [5–7]. These transforms are of interest to the Adiabatic Quantum Computing research community [12] since this community typically converts their optimization problems into Quadratic Unconstrained Boolean Optimization problems (QUBO). This idea should be familiar to computer scientists, since "transforms" are closely related to the "reductions" that are used to prove NP-completeness by mapping one problem class onto another problem class [11]. A keystone reduction in NP-completeness theory is the reduction that maps all satisfiability problems (SAT) onto the class of MAX-3SAT problems. Ansotegui et al. [2] have proposed new random SAT generators based on the power-law distribution, which is followed by industrial SAT instances [1]. We propose here an alternative way of generating random industrial-like MAX-SAT instances.

We stress that transforms and reductions can *smash* nonlinearity. Reductions and transforms can take problem instances with exponential nonlinearity as input and output a problem instance with dramatically bounded nonlinearity. Given any pseudo-Boolean function, transforms can output a quadratic form for QUBO instances or a cubic form for MAX-3SAT instances. This is a powerful idea, but it has not been used often in the evolutionary computation community, despite Kaufman's theoretical work suggesting that DNA may be characterized by k-bounded interactions [21].

Section 2 introduce a novel way to generate random MAX-3SAT instances with low nonlinearity, creating problems that are closer to real-world applications. This new method also builds on the idea that most real-world applications are constructed using either transforms or higher-order logical expressions and logical constraints. In Sect. 3 we introduce a new way to "weight" MAX-SAT instances that preserves lower nonlinearity. Surprisingly, these MAX-3SAT problems with randomized weights superficially resemble random NK landscapes.

In Sect. 4 we discuss Partition Crossover and its properties. Partition crossover induces lattices of local optima and sub-local optima. A **sub-local optima** is a solution that is locally optima in a hypercube subspace of the full search space. While one could define numerous hyperplane subspaces, we limit our attention to hyperplane subspaces that are induced by recombination events where the parents are themselves local optima in the full search space.

In Sect. 5 we enumerate all of the local optima for small weighted MAX-3SAT instances. We also enumerate all lattices that can be obtained by recombining all possible pairs of local optima. We find that these weighted MAX-3SAT instances induce more lattices than similar NK landscapes and thus appear to be more "searchable" than NK landscapes using Partition Crossover.

2   A New Construction for MAX-3SAT Problems

It is commonly assumed that real-world MAX-SAT applications have significantly lower degrees of nonlinearity than randomly generated MAX-SAT benchmarks [8,20]. It is often suggested that this is true because real-world problems have regular structure and repeating logical constraints. However, it is common to use "reductions" to convert a general satisfiability problem (SAT) to MAX-SAT. These reductions are a type of transform that reduces nonlinearity by adding additional auxiliary variables [6,7,11].

Thus, do industrial applications have low nonlinearity because of their regular structure, or because transforms have been used to convert these problems in MAX-SAT instances? We will show that the use of transforms can also generate random MAX-3SAT instances with surprisingly low nonlinearity.

The classic textbook on "Algorithms" by Cormen et al. [11] reduces the following 3 variable expression to Conjunctive Normal Form (CNF):

$$(y1 \iff (y2 \wedge \neg x2))$$

This expression is exactly as it appears in Cormen's reduction example. Variable $x2$ is a variable in the original SAT expression. Variables $y1$ and $y2$ are auxiliary variables used to convert the SAT expression into a binary tree. This 3 variable expression is then converted into the following four CNF clauses:

$$(\neg y1 \vee \neg y2 \vee \neg x2) \wedge (\neg y1 \vee y2 \vee \neg x2) \wedge (\neg y1 \vee y2 \vee x2) \wedge (y1 \vee \neg y2 \vee x2)$$

Note that each clause is made false by exactly one (distinct) assignment so that only one clause at a time is false. **This generalizes to all possible sets of 4 distinct CNF clauses over the same 3 variables.** The four assignments which make exactly one of the above CNF clauses false in our example are (in order): 111, 101, 100, 010. Because they share the same variables, these clauses combine into one subfunction (mapping the bit strings to integers), which we write in a single line as follows:

$$f_i(y1, y2, x2) = \overbrace{4}^{000}, \overbrace{4}^{001}, \overbrace{3}^{010}, \overbrace{4}^{011}, \overbrace{3}^{100}, \overbrace{3}^{101}, \overbrace{4}^{110}, \overbrace{3}^{111}$$

This can be shifted by a constant to yield

$$f_i(y1, y2, x2) = 1, 1, 0, 1, 0, 0, 1, 0$$

This is now a standard truth table over the 3 variables and gives us a unique way to identify all possible such constructions. In general, there are $\binom{2^k}{c}$ logical expressions over k variables assuming there are c clauses, and for every CNF clause there is 1 way in which is it false.

For MAX-3SAT, $k = 3$ and $c = 4$ when there are 4 clauses. Thus we have exactly $\binom{8}{4} = 70$ logical expressions. We next compute the Walsh transform of these functions (which is identical to the Hadamard or discrete square-wave Fourier Transform). Given a Boolean function $f : \{0,1\}^n \rightarrow \{0,1\}$, the ith Walsh coefficient for f is defined as:

$$w_i^{(f)} = \frac{1}{2^n} \sum_{x \in \{0,1\}^n} f(x)(-1)^{x \cdot i}, \tag{1}$$

where $x \cdot i$ denotes the dot product of the binary string x and the binary string represented by integer i.

Lemma 1. *Let $f : \{0,1\}^n \to \{0,1\}$ be a Boolean function with c unsatisfiable input combinations (zeroes). Let $\bar{f}$ represent its complement. Then the Walsh coefficients of f and $\bar{f}$ are related as follows:*

$$w_i^{(\bar{f})} = -w_i^{(f)} \quad \forall i \neq 0, \tag{2}$$

$$w_0^{(\bar{f})} = 1 - w_0^{(f)} = \frac{c}{2^n}. \tag{3}$$

Proof. Using Eq. (1) we get:

$$w_i^{(\bar{f})} = \frac{1}{2^n} \sum_{x \in \{0,1\}^n} (1 - f(x))(-1)^{x \cdot i}$$

$$= \frac{1}{2^n} \sum_{x \in \{0,1\}^n} (-1)^{x \cdot i} - \frac{1}{2^n} \sum_{x \in \{0,1\}^n} f(x)(-1)^{x \cdot i}$$

$$= \delta_i^0 - w_i^{(f)}$$

where we used the fact that $\sum_{x \in \{0,1\}^n} (-1)^{x \cdot i} = 2^n \delta_i^0$ (see [27]). We complete the proof with the following expression:

$$w_0^{(\bar{f})} = \frac{1}{2^n} \sum_{x \in \{0,1\}^n} (1 - f(x))(-1)^0 = \frac{c}{2^n}$$

□

According to Lemma 1, if we have a function with $n = 3$ and $c = 4$ unsatisfying assignments, $w_0^{(\bar{f})} = 1 - w_0^{(f)} = 4/8 = 0.5$.

This lemma tells us that to evaluate the nonlinearity of functions with $k = 3$ and $c = 4$ we only need to consider 35 cases. The following is a sample of 7 of the 35 cases, as well as the complement of the penultimate string (as an example). The coefficients and polynomials were generated using a Walsh transform.

	w_0	w_1	w_2	w_3	w_4	w_5	w_6	w_7
0 0 0 0 1 1 1 1	0.5	0	0	0	−0.5	0	0	0
0 0 0 1 0 1 1 1	0.5	−0.25	−0.25	0	−0.25	0	0	0.25
0 0 1 0 0 1 1 1	0.5	0	−0.25	−0.25	−0.25	0.25	0	0
0 1 0 0 0 1 1 1	0.5	−0.25	0	−0.25	−0.25	0	0.25	0
0 1 0 0 1 0 1 1	0.5	0	0	0	−0.25	−0.25	0.25	−0.25
0 1 0 0 1 1 0 1	0.5	−0.25	0.25	0	−0.25	0	0	−0.25
0 1 0 0 1 1 1 0	0.5	0	0.25	−0.25	−0.25	−0.25	0	0
1 0 1 1 0 0 0 1	0.5	0	−0.25	0.25	0.25	0.25	0	0

Note that w_0 is a constant, the coefficients w_3, w_5, w_6 are quadratic and w_7 is cubic. The other coefficients are linear. It is easy to prove that for every coefficient w_i with $i \neq 0$, 36 of the 70 entries are zero[1]. This also removes more than half of the cubic coefficients, w_7.

What is striking about this simple result is the fact that these random MAX-3SAT constructions display very low nonlinearity.

By contrast, here are the Walsh coefficients for a single example clause.

								w_0	w_1	w_2	w_3	w_4	w_5	w_6	w_7
1	1	1	0	1	1	1	1	0.875	0.125	0.125	−0.125	−0.125	0.125	0.125	−0.125

By enumeration, every possible single clause has coefficients with weights $w_0 = 7/8 = 0.875$ and $w_i = \pm 1/8 = \pm 0.125$. (The signs are the negation of one column of the corresponding 8 by 8 Walsh matrix.)

Next, assume we generate a random MAX-3SAT instance by generating a random 3-variable SAT expression and then we convert it into 4 CNF clauses. We will refer to these as Reduction-based MAX-3SAT instances, or *Reduced MAX-3SAT* to be more concise. Assume this instance also has m clauses. In expectation, the majority of cubic terms in a Reduced MAX-3SAT instance will cancel. First, for every 4 CNF clauses that touch the same 3 variables, there is only one possible cubic term. Second, with a probability of 36/70 the possible cubic term of the corresponding Walsh polynomial over the 4 CNF clauses will be zero. If $m = 4n$ the number of cubic terms is approximately $n/2$ instead of the $4n$ we would expect for m completely random clauses. *This represents an 8 fold decrease in cubic terms.* In addition, the Reduced MAX-3SAT instance will have approximately $3n$ quadratic coefficients compared to the approximately $12n$ quadratic terms we would expect for m completely random clauses. And, again with probability 36/70 each possible quadratic term of the corresponding Walsh polynomial over the 4 CNF clauses will be zero. Thus, by generating random SAT expressions (truth tables) over 3 variables, and then converting these into MAX-3SAT expressions, the level of nonlinearity is dramatically reduced.

We will express these arguments in terms of the expected number of Walsh Coefficients Per Clause (WPC). Let v denote the clause-to-variable ratio of a MAX-SAT instance so that $m = vn$.

Theorem 1. *Assume that all variable interactions are unique for each clause or for each Reduced MAX-3SAT subfunction and all linear coefficients are non-zero. Assume $m = vn$. Then the Walsh Coefficients Per Clause (WPC) for Reduced MAX-3SAT instances is $WPC < 1/v + 0.5$. The Walsh Coefficients Per Clause (WPC) for standard random MAX-3SAT instances is $WPC = 4 + 1/v$.*

Proof. There are n linear coefficients. If $m = vn$ the WPC for linear coefficients is $n/m = 1/v$. We next use counting arguments from lemma 1. For Reduced MAX-SAT, there are only 3 quadratic coefficients for every 4 CNF clauses. More than

---

[1] In general, each Walsh coefficient except w_0 is zero in $\binom{2^{k-1}}{c/2}^2$ out of $\binom{2^k}{c}$ functions.

half of these are zero (36/70 to be precise) in expectation. Thus, the number of quadratic terms is bounded by $(3/4m * 0.5)$ and $WPC < 3/8$ for the quadratic coefficients. Finally, the number of cubic terms is bounded by $m/4 * 0.5$ and $WPC < 1/8$. Since we assume that all variable interactions per clause are unique, a normal random MAX-3SAT clause has 3 quadratic and 1 cubic terms per clause, all with weight ± 0.125 and $WPC = 1/v + 4$.   □

Of course, all variable interactions are not unique, but if $m = O(n)$, the variable interactions of random instances have a high probability of being unique. For quadratic interactions, we are sampling $O(n)$ elements from $O(n^2)$ choices, and for cubic interactions, we are sampling $O(n)$ elements from $O(n^3)$ choices.

If $m = 4n$ (which is close to the random MAX-3SAT phase transition [22]) standard random MAX-3SAT instance will have an expected approximate WPC = 4.25. However, for a Reduced MAX-3SAT instance the expected approximate WPC = 0.75. There is a dramatic difference in nonlinearity.

The Reduced MAX-3SAT instances are compared with 14 large industrial problems from the MAX-SAT 2012 challenge[2] to determine their level of non-linearity. Problem size ranged from 247,943 to 2,766,036 variables. These real-world problems have a median WPC of just 1.17 Walsh coefficients per clause (for the mem-control problem) [17]. Table 1 shows that the WPC ranges from 1.69 (the maximum) to 0.73 (the minimum). $WPC = 1.26$ is the second largest and $WPC = 0.978$ is the second smallest.

Table 1. WPC of five instances of the MAX-SAT 2012 competition.

Problem Name	b15	fpu	mem control	wb-46	rsd-KES
WPC	1.69	1.26	1.17	0.978	0.73

For these industrial problems, the WPC is between 0.978 and 1.26 for 12 of the 14 industrial problems. As can be seen, our Reduced MAX-3SAT problems have slightly lower WPC than these industrial problems. But the WPC of our Reduced MAX-3SAT problems is much closer to these industrial problems than the WPC of standard random MAX-3SAT instances, where the typical WPC is greater than 4.1 for problems near the phase transition.

Our Reduced MAX-3SAT problems could be preprocessed to remove some variables and clauses. Some output vectors (truth tables) have lower nonlinearity because they express simple constraints on the MAX-3SAT constructions.

For example, if the variables are x_1, x_2, x_3 then the following linear constraints represents the truth table where $x_3 = 1$ (x_3 is true).

	w_0	w_1	w_2	w_3	w_4	w_5	w_6	w_7
0 0 0 0 1 1 1 1	0.5	0	0	0	−0.5	0	0	0

---

[2] http://MAX-SAT.ia.udl.cat/.

Thus, the resulting Walsh polynomial only has one linear coefficient. Variable x_3 could be removed, as well as the associated clauses that are satisfied by x_3 by standard unit propagation (which would increase the WPC). This would also reduce the size of clauses that reference x_3. This kind of preprocessing is normal for real-world applications [14].

3   Weighted MAX-3SAT

Weighting MAX-SAT instances has various purposes. Weights may be used to distinguish between hard constraints that are important (for clauses that must be satisfied) and soft constraints (clauses one would prefer to satisfy, but satisfaction is not a hard constraint) [23].

For MAX-kSAT instances, local optima may not be well-defined because of very large plateaus in the search space [15,19]. We use weighted MAX-SAT instances to break ties when possible, and thus break up the potentially large plateaus that commonly occur in the landscape of MAX-kSAT instances.

We have as an additional goal to minimize nonlinearity to the degree possible. We introduce a new way to add weights so that the weights do not increase the number of 3-way cubic interactions in the Walsh polynomial.

First, consider the following 4 MAX-SAT clauses over the same 3 variables and its Walsh transform.

$f(x_1, x_2, x_3) =$

	1	0	1	1	0	0	0	1
Walsh	w_0	w_1	w_2	w_3	w_4	w_5	w_6	w_7
	0.5	0	-0.25	0.25	0.25	0.25	0	0

Next, the UNSAT assignments are given a very small distinct weight, while the SAT assignments are all weighted by 4, so that the global maximum returns the total of clauses satisfied. Next, we again compute the Walsh coefficients.

$f(x_1, x_2, x_3) =$

	4	0.12	4	4	0.13	0.11	0.14	4
Walsh	w_0	w_1	w_2	w_3	w_4	w_5	w_6	w_7
	2.0625	0.005	-0.9725	0.97	0.9675	0.965	0.0025	0.0

Note that the cubic term is still zero: $w_7 = w_{111} = 0.0$. This is because the Walsh signs cause the weight to cancel: $w_7 = 0.14 + 0.11 - 0.12 - 0.13 = 0$.

The weights can be scaled to be arbitrarily small, while also ensuring that any ranking ordering of solutions can be preserved.

$$w_7 = w_{111} = 0.00014 + 0.00011 - 0.00012 - 0.00013 = 0.0.$$

Fortunately, even very small weights are sufficient to disrupt plateaus and allow smaller distinct local optima (or smaller plateaus) to form. Note that these weights can be quasi-random except they must sum to zero. In practice, three weights can be random, and the fourth weight is then determined.

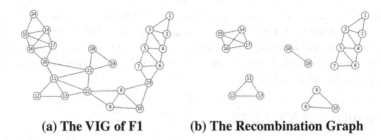

(a) The VIG of F1   (b) The Recombination Graph

Fig. 1. Variable interaction graph (VIG) for example F1 as well as the decomposed recombination graph for P1 and P2.

4   Partition Crossover, Local Optima and Lattices

Partition crossover is a deterministic recombination operator that breaks the parents into linearly separable recombining components [28]. In general, this is only possible if the evaluation function has limited nonlinearity. But many NP-hard problems do display limited nonlinearity, including MAX-kSAT [9,10], NK landscapes, QUBO, and the Traveling Salesman Problem [18].

We will use a MAX-3SAT example borrowed from Dunton [13]. The following function F1 has $m = 18$ subfunctions and $n = 24$ variables. The subfunction are labeled from f_a to f_z (but using only 18 characters). The variables are labeled with integers from 1 to 24, where integer i denotes variable x_i.

F1: f_a: 1 2 3 f_b: 2 3 4 f_c: 3 4 5 f_d: 4 5 6 f_e: 5 6 7
    f_r: 6 7 23 f_h: 8 9 22 f_i: 8 10 23 f_j: 11 12 13 f_s: 8 9 10
    f_k: 11 13 22 f_l: 11 20 21 f_m: 11 21 22 f_n: 14 15 24
    f_o: 14 16 17 f_p: 15 16 17 f_q: 16 17 20 f_z: 18 19 21

Figure 1 shows the variable interaction graph (VIG) for function F1. A VIG is a graph where vertices represent variables, and there is an edge $e(v_i, v_j)$ if and only if there is a nonlinear interaction between vertex v_i and v_j.

Figure 1 is based on a specific MAX-3SAT instance and we will assume that none of the Walsh coefficients are zero. In practice, it is important to use Walsh coefficients when computing the VIG for Reduced MAX-3SAT functions. Removing the canceled interactions can reduce the number of edges in the VIG by 50%. When $m = O(n)$ the resulting VIG has $O(n)$ edges, and the total number of coefficients in the resulting Walsh polynomial is $O(n)$. Thus, the VIG is computed in $O(n)$ time using the Walsh transform. The VIG is only computed once.

Let P1 denote Parent 1 and P2 denote Parent 2. Furthermore, assume P1 and P2 are local optima under a Hamming bit-flip neighborhood[3]. The first step of Partition Crossover for k-bounded pseudo-Boolean function is to remove vertices from the VIG that corresponds to variables where P1 and P2 have a

---

[3] However, any neighborhood local search operator will suffice.

shared assignment (both 1 or both 0). Thus, the variables that remain in the VIG are always complements (0 and 1, or 1 and 0). After variables with the same assignment are removed, the VIG is often decomposed into q connected subgraphs. Let the two parent solutions P1 and P2 be:

$$P1 = 01000\ 01100\ 10000\ 01101\ 1111$$

$$P2 = 10111\ 10011\ 01111\ 10011\ 1111$$

These two solutions share only the last 5 bits in common. After vertices 20 to 24 (x_{20} to x_{24}) are deleted, there are five connected subgraphs. Figure 1b shows the shattered graph for function F1, and parents P1 and P2. This "shattered graph" represents the "recombination graph."

Assuming that the VIG is shattered into q connected subgraphs, the evaluation function $f(x)$ can (temporarily) be decomposed into a set of new subfunctions $g_1(x) \cdots g_q(x)$. Let x' denote any string where the shared bits in P1 and P2 are fixed. Then

$$f(x') = g(x') = \sum_{i=1}^{q} g_i(x')$$

The new subfunctions $g_i(x')$ decomposes both the set of subfunctions as well as the set of variables. Logically it follows that if the VIG has been shattered, this must decompose both the variables and the subfunctions. For example, in the shattered recombination graph in Fig. 1, variables 14, 15, 16, 17 are associated with the original subfunctions f_n, f_o, f_p, f_q while variables 11, 12, 13 are associated with the original subfunctions f_j, f_k, f_l, f_m.

4.1 Lattices and Linear Equations

There exists a simple linear equation that can be used to evaluate all of the children of parents P1 and P2 under Partition Crossover. This can be applied to all k-bounded pseudo-Boolean functions, as well as Traveling Salesman Problems when using Partition Crossover.

Let q denote the number of recombining components. Let b denote an auxiliary binary string of length q. Let $b_i = 0$ denote that the offspring inherits bits from parent 2 in the ith recombining component. Similarly, $b_i = 1$ denotes that the offspring inherits bits from parent 1 in the ith recombining component. When $b = 1^q$ all bits are inherited from P1 (the offspring is a copy of P1). When $b = 0^q$ all bits are inherited from P2 (the offspring is a copy of P2).

Theorem 2 (Theorem 3.1 in [29]). Under Partition Crossover, all of the 2^q children of parents P1 and P2 can be evaluated using the following linear equation and the auxiliary bit function b. For any child x_c

$$f(x_c) = \alpha_0 + \sum_{i=1}^{q} \alpha_i b_i \tag{4}$$

where $\alpha_0 = f(P2)$ and $f(P1) = \alpha_0 + \sum_{i=1}^{q} \alpha_i$.

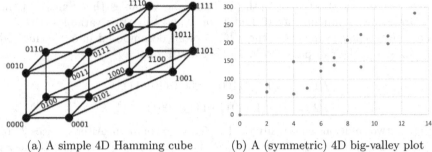

(a) A simple 4D Hamming cube   (b) A (symmetric) 4D big-valley plot

Fig. 2. The big-valley (fitness-distance correlation) plot shows the relative location of local optima where Hamming distance from the global optimum is plotted on the x-axis, and fitness is on the y-axis. But looking closely, the big-valley plot is also a 4D Hamming cube associated with 16 local optima generated by one Partition Crossover event. The big-valley distribution is plotted in terms of minimization instead of maximization.

Assume the objective function is a random NK landscape, and we discover a lattice with $q = 7$ recombining components, and the linear equation which is induced is given by

$$f(x) = \alpha_0 + 7b_1 - 143b_2 - 29b_3 + 52b_4 + 83b_5 - 16b_6 + 66b_7.$$

The bit mask returns the best possible offspring out of 128 possible children by selecting only the positive coefficients (assuming maximization). Thus the b vector for the best possible offspring is $b = 1001101$.

Note that the b vector defines a hypercube which in some sense *organizes* the offspring. This hypercube induces a non-planar *lattice*.

A **lattice** is a partially ordered set containing the children and parents of a Partition Crossover event. In some sense, a lattice is just a weighted Hamming cube. A lattice can be ordered by both fitness, as well as Hamming distance. All of these children and parents in a lattice are related in fitness via Equation (4). Tinos et al. [28] proved that if the parents are local optima, then all children of Partition Crossover must also be locally optima in the smallest hyperplane subspace that contains the parents. If all of the children are local optima in the search space, then all of these local optima have a fitness value determined by Equation (4). **Thus, the evaluations of local optima in the search are not arbitrary, but rather determined by the lattices to which they belong as well as by Equation** (4). This is true for all local optima that appear in at least one Partition Crossover lattice.

A 4D Hamming cube is shown in Fig. 2, as well as the big-valley (fitness-distance correlation) plot of local optima generated by single specific Partition Crossover event. A big-valley plot is a plot of local optima with fitness on the y-axis and Hamming distance (usually relative to the global optimum) on the x-axis. A big-valley plot is traditionally shown in terms of **minimization** as is the case here. Because Hamming distance is itself a linear calculation, there is also a linear equation function for Hamming distance with respect to the b vector used in Eq. 4. All of the 16 children shown in Fig. 2b are local optima. Careful inspection of Fig. 2b shows that these local optima also form a 4D hypercube, weighted by $2b_1 + 2b_2 + 4b_3 + 5b_4$ for Hamming distance relative to the best child, and $85b_1 + 59b_2 + 64b_3 + 75b_4$ for fitness (with respect to 0,0 for the best solution). Only complementary strings can regenerate the entire lattice. Complementary bit strings can be either far apart or close together in terms of fitness. All complementary offspring are equidistant from the evaluation $(f(P1)+f(P2))/2$ and any relative Hamming distance $(HD(P1)+HD(P2))/2$. This is not a coincidence, and it impacts the probability that Partition Crossover is able to find improving moves. This also illustrates why it is important to see the Hamming cube containing all of the children as a lattice. The lattice organizes the children in terms of both fitness and relative Hamming distance.

5   Enumeration Results

In this section, we look at the hypercube lattices associated with both NK land-scapes, as well as Weighted Reduced MAX-3SAT instances. We do this by enumerating all of the local optima in the search space for problems with $n = 30$ variables. After we enumerate all of the local optima in the search space, we then recombine all pairs of local optima in the search space. This yields all lattices that can be reached by recombining local optima. In the case of Weighted Reduced MAX-3SAT instances, local optima sometimes take the form of relatively small plateaus of solutions with equal fitness.

5.1   A Trick to Enumerate Local Optima

The enumeration of local optima often **does not** require enumeration of the entire search space. If we have the Walsh coefficients for a k-bounded pseudo-Boolean function, we also know which variables have nonlinear interactions.

Assume $n = 30$ bits (n can be arbitrary), and that we know that when starting from the string 0^{30} that an improving move can be obtained by flipping bit number 25. Thus 0^{30} is not a local optimum. Furthermore assume that we know that bit 25 interacts only with bits 23, 25, 29, 30. Until some bit is changed that has a nonlinear interaction with bit 25, any string (solution) that we visit cannot be a local optimum (because flipping bit 25 still produces an improving move). Thus we can identify hyperplanes of the search that can be ignored (and thus eliminated) because they do not contain a local optimum. In this way, all local optima can be found without enumerating the entire search space [16].

5.2   Empirical Results and Trends

Table 1 presents empirical data based on enumeration. We enumerated the local optima in 10 random NK landscapes and 10 Reduced MAX-3SAT instances.

There are $n = 30$ variables for all problems. By enumerating these smaller problems, we can answer questions that would be impossible otherwise.

The random NK landscapes and the Reduced MAX-3SAT instances used exactly the same random variable interaction graphs. Thus exactly the same random assignment of variables to subfunctions was used for both classes of problems. Any differences are **not** due to the variable interaction graph. All 8 values in the subfunctions of the NK landscape are random; for the Reduced MAX-3SAT instances 4 out of 8 values are a constant (when all 4 clauses are satisfied) while 4 values correspond to a different randomized penalty when one clause is not satisfied. Thus, differences should be due to the reduction in the nonlinearity in our construction of Reduced MAX-3SAT instances.

We note that only one of the 10 MAX-3SAT instances is satisfiable, and 9 instances are UNSAT. For 3 of the other 9 instances 119 out of 120 clauses were satisfiable. Next, 118 out of 120 clauses were satisfiable for 3 of 9 instances, and 117 out of 120 clauses were satisfiable for 3 instances. Keep in mind these are true global optima. This also likely means that **the phase transition for Reduced MAX-3SAT is different than that of standard random MAX-3SAT instances**.

All of our problems have a variable-to-clause ratio of 4.0. The phase transition for standard random MAX-3SAT instances is a 4.27 variable to clause ratio [22]. Thus, we had expected that more than one of our Reduced MAX-3SAT instances would also be satisfiable. We conjecture that Reduced MAX-3SAT has a phase transition that is less than 4.0, as it also the case with other real-like random instances [4]. Note that an instance with a global optimum of 119 out of 120 clauses could be made SAT by deleting just one clause.

For the random NK landscapes, we present average results for various metrics. For the Reduced MAX-3SAT instances we show a range of problem instances ranked by number of local optima (which corresponds exactly to the ranking by number of lattices). We show the averages for the various metrics, as well as the specific metrics for the problems with rank 1, 4, 7 and 10.

The metrics include, among others, the number of local optima and the number of global solutions. For MAX-3SAT in particular there can be numerous ties for the global optimum, and the global optima was always part of a plateau. There were more globally optimal solutions when the evaluation of the global optima was 117 or 118. Thus, when the problems were more constrained (and the global optimum satisfied fewer clauses) there were larger plateaus and multiple plateaus representing multiple global optima that were easier to find. This also provides additional evidence that the phase transition for Reduced MAX-3SAT is likely different than that of standard random MAX-3SAT instances.

One thing that is notable is that Reduced MAX-3SAT instances have far more local optima than NK landscape where $k = 3$ ($K = 2$). These local optima also induce more tunnels between optima, and more pathways to the global optima. The number of lattices that reach a globally optimal solution is surprisingly high, but there is great variance between samples.

Table 2. Empirical data collected by enumerating all local optima for N = 30. All numbers are averages of 10 instances. The Random NK landscapes results are averages. The symbol # denotes a count in this table.

	NK	Reduced MAX-3SAT Instances				
	Random	MAX-3SAT	Rank 1	Rank 4	Rank 7	Rank 10
	Mean	Mean	Seed 1	Seed 9	Seed 6	Seed 4
Local Optima #	485	1984	659	1597	2434	3095
Size 4 lattices	27,348	912,362	54,131	435,155	1,002,621	3,216,771
Size 8 lattices	4,280	492,550	11,144	182,632	390,446	2,538,847
Size 16 lattices	284	192,817	771	58,730	107,157	1,227,573
Size 32 lattices	1	53,119	7	14,767	17,820	392,258
Size 64 lattices	0	9,056	0	2,197	1,282	75,693
Size 128 lattices	0	768	0	98	16	6,940
Size 256 lattices	0	21	0	0	0	208
Size 512 lattices	0	0	0	0	0	0
Total Lattices	31,915	1,662,694	66,053	693,579	1,539,342	7,458,290
Crossover Success	21%	51%	30.46%	54.42%	51.98%	75.78%
Sub-Opt Children	26%	11.6%	8.58%	12.83%	12.62%	15.91%
Global Solutions	1.2	12.6	4	108	54	444
Clauses Satisfied	na	118.2	119	118	120	117
Global Lattices%	1.79%	9.4%	1.55%	13.55%	20.32%	20.61%

An additional metric is the crossover success frequency: when PX is used to recombine two local optima, how often does that recombination produce offspring? The crossover success frequency is lower for random NK landscapes: 21 percent of the time PX produces offspring. For Reduced MAX-3SAT problems the crossover success frequency is 51 percent (Table 2).

We also report the total number of lattices found as well as how many lattices were found of specific size (from size 4 to 256, corresponding to $q=2$ to $q=8$ recombining components). On average there were 1,662,694 lattices found for Reduced MAX-3SAT instances. There were on average only 31,915 lattices found for random NK landscapes. The number of lattices found is related to both the Crossover Success rate and the number of local optima. As the number of lattices increases, the size of the largest lattices also increases. The Reduced MAX-3SAT instances sometimes included lattices of sizes 128 and 256. There were no lattices of size larger than 32 for random NK landscapes.

We also report the number of sub-optima (children of PX that were not locally optimal). This was highest for random NK landscapes at 26 percent, so that 74 percent of all children were also local optima in the full space. For Weighted Reduced MAX-3SAT 88 percent of all children were also local optima.

Of course, the local optima of Weighted Reduced MAX-3SAT problems are often small plateaus.

And finally, what percentage of the lattices included a globally optimal solution? For Weighted Reduced MAX-3SAT this was 9.4% percent. This sounds very encouraging, but this number should be taken with a grain of salt. Not every lattice that contains a globally optimal solution is equally easy to discover, and for some lattices of size 4, reaching the global optimum could be difficult. Furthermore, if we only look at the MAX-3SAT with 119/120 clauses satisfied, the percentage is only 3.14%. For these instances, the global plateaus were much smaller and hard to find. On the other hand, for the single SAT instance (120/120 clauses satisfied), the percentage is 20.32%. A better understanding of the phase transition for Reduced MAX-3SAT problems should shed additional light on the question of problem difficulty.

Overall, this empirical data suggests that Reduced MAX-3SAT instances can often be searched effectively using Partition Crossover as an operator. Reduced MAX-3SAT instances also seem much easier to search than random NK landscapes. More work is needed to understand whether problems drawn from closer to a phase transition might be more difficult than the problems we have sampled.

6   Conclusions

This paper explores the nonlinearity of MAX-SAT instances, and introduces weighted and unweighted Reduced MAX-3SAT problems which display low nonlinearity; low linearity is more typically associated with industrial MAX-SAT applications. Standard randomly generated MAX-3SAT instances have a Walsh coefficient per clause count that is approximately 5 times larger than unweighted Reduced MAX-3SAT ($4.25/0.75 > 5$); all of these additional coefficients are quadratic and cubic variable interactions. A new way to create Weighted Reduced MAX-3SAT problems is also described which does not introduce any additional cubic variable interactions.

Finally, we look at how Partition Crossover can be applied to Weighted Reduced MAX-3SAT problems. Partition crossover induces hypercube lattices over subsets of local optima which are found when Partition Crossover is used to recombine two locally optimal solutions. In this paper, the local optima are induced by a single-bit flip neighborhood. By enumerating local optima for instances with $n = 30$ variables, we found that Partition Crossover is highly successful at "tunneling" between local optima. More work is needed to understand the phase transition of Reduced MAX-3SAT problems and how this might impact problem difficulty.

Acknowledgements. This work was supported by an National Science Foundation (NSF) grant to D. Whitley, Award Number:1908866. This work was also partially funded by *Universidad de Málaga, Ministerio de Ciencia, Innovación y Universidades del Gobierno de España* under grant PID 2020-116727RB-I00, and by EU Horizon 2020 research and innovative program (grant 952215, TAILOR ICT-48 network).

References

1. Ansótegui, C., Bonet, M.L., Levy, J.: On the structure of industrial SAT instances. In: Gent, I.P. (ed.) Principles and Practice of Constraint Programming - CP 2009. Lecture Notes in Computer Science, vol. 5732, pp. 127–141. Springer, Berlin (2009). https://doi.org/10.1007/978-3-642-04244-7_13
2. Ansótegui, C., Bonet, M.L., Levy, J.: Towards industrial-like random SAT instances. In: IJCAI, vol. 9, pp. 387–392 (2009)
3. Balint, A., Fröhlich, A.: Improving stochastic local search for SAT with a new probability distribution. In: Strichman, O., Szeider, S. (eds.) Theory and Applications of Satisfiability Testing - SAT 2010. Lecture Notes in Computer Science, vol. 6175, pp. 10–15. Springer, Berlin (2010). https://doi.org/10.1007/978-3-642-14186-7_3
4. Bläsius, T., Friedrich, T., Sutton, A.M.: On the empirical time complexity of scale-free 3-SAT at the phase transition. In: Vojnar, T., Zhang, L. (eds.) Tools and Algorithms for the Construction and Analysis of Systems. Lecture Notes in Computer Science(), vol. 11427, pp. 117–134. Springer, Cham (2019). https://doi.org/10.1007/978-3-030-17462-0_7
5. Boros, E., Crama, Y., Rodriquez-Heck, E.: Compact quadratizations for pseudo-Boolean functions. J. Comb. Optim. **39**, 687–707 (2020)
6. Boros, E., Gruber, A.: On quadratization of pseudo-Boolean functions (2014). arXiv:1404.6538
7. Boros, E., Hammer, P.: Pseudo-Boolean optimization. Discr. Appl. Math. **123**(1), 155–225 (2002)
8. Braunstein, A., Mézard, M., Zecchina, R.: Survey propagation: an algorithm for satisfiability. Random Struct. Algorithms **27**(2), 201–226 (2005)
9. Chen, W., Whitley, D.: Decomposing SAT instances MAX-kSAT with pseudo backbones. In: European Conference on Evolutionary Computation in Combinatorial Optimization, LNCS, vol. 10197, pp. 75–90. Springer, Cham (2017)
10. Chen, W., Whitley, D., Tinós, R., Chicano, F.: Tunneling between Plateaus: improving on a state-of-the-art MAXSAT solver using partition crossover. In: GECCO Genetic and Evolutionay Computation Conference, pp. 921–928. ACM (2018)
11. Cormen, T., Leiserson, C., Rivest, R.: Introduction to Algorithms. McGraw Hill, New York (1990)
12. Dattani, N.: Quadratization in discrete optimization and quantum mechanics (2019). ArXiv:1901.04405
13. Dunton, P., Whitley, D.: Reducing the cost of partition crossover on large MAXSAT problems: the PX-preprocessor. In: Genetic and Evolutionary Computation Conference (GECCO-2022), pp. 694–702 (2022)
14. Eén, N., Biere, A.: Effective preprocessing in SAT through variable and clause elimination. In: Bacchus, F., Walsh, T. (eds.) Theory and Applications of Satisfiability Testing. Lecture Notes in Computer Science, vol. 3569, pp. 61–75. Springer, Berlin (2006). https://doi.org/10.1007/11499107_5
15. Frank, J., Cheeseman, P., Stutz, J.: When gravity fails: local search topology. J. Artif. Intell. Res. **7**, 249–281 (1997)
16. Goldman, B., Punch, W.: Hyperplane elimination for quickly enumerating local optima. In: Chicano, F., Hu, B., Garcia-Sanchez, P. (eds.) Evolutionary Computation in Combinatorial Optimization. Lecture Notes in Computer Science(), vol. 9595, pp. 154–169. Springer, Cham (2016). https://doi.org/10.1007/978-3-319-30698-8_11

17. Hains, D., Whitley, D., , Howe, A., Chen, W.: Hyperplane initialized local search for MAXSAT. In: Proceedings of GECCO'2009, pp. 805–812 (2013)
18. Helsgaun, K.: DIMACS TSP challenge results: current best tours found by LKH (2013). http://www.akira.ruc.dk/keld/research/LKH/DIMACSresults.html. Accessed 24 Nov 2013
19. Hoos, H.H.: On the run-time behaviour of stochastic local search algorithms for SAT. In: Proceedings of AAAI, pp. 661–666 (1999)
20. Hoos, H., Stützle, T.: Stochastic Local Search: Foundations and Applications. Morgan Kaufman, Massachusetts (2004)
21. Kauffman, S.: The Origins of Order. Oxford Press, Oxford (1993)
22. Kirkpatrick, S., Selman, B.: Critical behavior in the satisfiability of random Boolean expressions. Science **264**, 1297–1301 (1994)
23. Li, C., Manya, F.: MAXSAT, hard and soft constraints. In: Handbook of Satisfiability. IOS Press, Amsterdam (2021)
24. Lin, C., Wei, W., Zhang, H.: Combining adaptive noise and look-ahead in local search for SAT. In: Marques-Silva, J., Sakallah, K.A. (eds.) Theory and Applications of Satisfiability Testing - SAT 2007. Lecture Notes in Computer Science, vol. 4501, pp. 121–133. Springer, Berlin (2007). https://doi.org/10.1007/978-3-540-72788-0_15
25. Selman, B., Levesque, H., Mitchell, D.: A new method for solving hard satisfiability problems. In: AAAI, pp. 44–446 (1992)
26. Selman, B., Kautz, H., Cohen, B.: Local search strategies for satisfiability testing. In: Johnson, D.S., Trick, M.A. (eds.) DIMACS Series in Discrete Mathematics and Theoretical Computer Science, vol. 26. AMS (1996)
27. Sutton, A.M., Whitley, L.D., Howe, A.E.: A polynomial time computation of the exact correlation structure of k-satisfiability landscapes. In: GECCO 2009, pp. 365–372. ACM (2009). https://doi.org/10.1145/1569901.1569952
28. Tinós, R., Whitley, D., Chicano, F.: Partition crossover for Pseudo-Boolean optimization. In: Foundations of Genetic Algorithms, (FOGA-15), pp. 137–149 (2015)
29. Whitley, D., Ochoa, G., Chicano, F.: Partition crossover can linearize local optima lattices of k-bounded Pseudo-Boolean functions. In: Foundations of Genetic Algorithms. ACM (2023)

Where the Really Hard Quadratic Assignment Problems Are: The QAP-SAT Instances

Sébastien Verel[1(✉)], Sarah L. Thomson[2], and Omar Rifki[1]

[1] LISIC, Université du Littoral Côte d'Opale (ULCO), Boulogne-sur-Mer, France
{verel,omar.rifki}@univ-littoral.fr
[2] Edinburgh Napier University, Scotland, UK
S.Thomson4@napier.ac.uk

Abstract. The Quadratic Assignment Problem (QAP) is one of the major domains in the field of evolutionary computation, and more widely in combinatorial optimization. This paper studies the phase transition of the QAP, which can be described as a dramatic change in the problem's computational complexity and satisfiability, within a narrow range of the problem parameters. To approach this phenomenon, we introduce a new QAP-SAT design of the initial problem based on submodularity to capture its difficulty with new features. This decomposition is studied experimentally using branch-and-bound and tabu search solvers. A phase transition parameter is then proposed. The critical parameter of phase transition satisfaction and that of the solving effort are shown to be highly correlated for tabu search, thus allowing the prediction of difficult instances.

Keywords: Quadratic Assignment Problem · Phase transition

1 Introduction

The Quadratic Assignment Problem (QAP) has held major importance within evolutionary computation research for decades [1–3,31]. Given a matrix of flow between abstract objects, and a distance between positions, the goal of QAP [19] is to find an assignment of objects to positions in order to minimize the sum of costs, *i.e.* the product of flow and distance, between all possible pairs of objects. The QAP is often considered one of the most difficult problems in the NP-hard class with many applications [22]. Notice that Traveling Salesperson Problem (TSP) is a special case of QAP with a dedicated flow matrix. In order to understand and improve optimization algorithms design, a large number of QAP instances with relevant properties have been proposed in the literature (see more details in Sect. 2).

Coming from statistical physics, the notion of a phase transition is also an important property in combinatorial optimization and for decision problems.

© The Author(s), under exclusive license to Springer Nature Switzerland AG 2024
T. Stützle and M. Wagner (Eds.): EvoCOP 2024, LNCS 14632, pp. 129–145, 2024.
https://doi.org/10.1007/978-3-031-57712-3_9

Phase transition is a phenomenon related to the rapid change around a critical value of an order parameter of the probability that a random instance is satisfiable. From the seminal work on SAT problems [7], this property has been shown in a large number of decision and combinatorial problems [4,17] (TSP, constraint problems, etc.). The phase transition is also associated with problem difficulty which also changes around the point of the phase transition. Indeed, problem instances defined around the critical value of phase transition between feasible and unfeasible are often considered as the most challenging, and the most interesting instances to solve. However, except for highly generic assignment problems [5] different from classical QAP, or special cases of QAP such as TSP [15], to our best knowledge no phase transition properties have been shown for QAP. One objective of this work is to show a phase transition in the "pure" QAP problem. To show the phase transition phenomenon, we propose new QAP instances with tunable difficulty: the QAP-SAT, based on submodularity (clauses) similar to the notion of a clause in SAT problem.

Phase Transition, and Problem Difficulty. Phase transition is a phenomenon that appears in randomly generated instances of intractable decision problems. It links the computational complexity to the satisfiability of the instances, such that a sudden change in the satisfiability happens in a narrow range of the instance parameters, summarized by an order parameter. The passage from easily solvable satisfiable problem instances to easily checked unsatisfiable problem instances can be seen. At the transition between both regions, where the order parameter crosses the critical value, hard random instances can be found. As this critical value has been shown to be independent of the solving algorithms, hard and easy instances can thus be located with precision, and subsequently used for fair benchmarking, algorithm selection, and configuration. The first applications on NP-complete problems date back to [7]. From this point, phase transition has been observed on most of the famously known NP-complete problems, such as the satisfiability problem [7,14,24], the traveling salesman problem [7,15], the graph coloring problem [7], the 0–1 knapsack problem [37], the minimum vertex cover problem [36], and so on.

Another way of characterizing the inherent structure of the search space is fitness landscape analysis: this provides a number of features. These landscape metrics happen to be valuable for gaining insight into the performance of algorithms on a given problem instance, thus relating to problem difficulty. We discuss landscape analysis for QAP in Sect. 2.3. For a general introduction to the topic of phase transition and problem difficulty, see [21,28]. It should be also mentioned that phase transition on random combinatorial optimization problems can be seen through the prism of statistical physics of disordered systems [4,17]. Models such as the spin glass model are used to exhibit the change in behavior of the problem satisfiability, for example, [25] for K-SAT.

The remaining of the paper is as follows. Section 2 presents the QAP, its formulation, famous benchmark datasets, and its problem features. Section 3 introduces the QAP-SAT decomposition. The experimental study is detailed in Sect. 4. Section 5 concludes the paper.

2 Quadratic Assignment Problems

2.1 Definition of QAP

The Quadratic Assignment Problem (QAP) [19] is a minimization assignment problem where the search space $\mathcal{S}_n$ is the symmetric group of dimension n, *i.e.* the set of permutations of size n. The QAP fitness function is defined as follows:

$$\forall \sigma \in \mathcal{S}_n, \; Q_{A,B}(\sigma) = \sum_{i=1}^{n} \sum_{j=1}^{n} A_{ij} B_{\sigma_i \sigma_j}$$

where A, and B are square matrices of real numbers of dimension $n \times n$. Usually, A is called a flow matrix. A_{ij} represents the flow (cost) between two abstract objects i, and j. B matrix is called distance matrix. B_{ij} represents the distance (cost) between positions i, and j. The objective function has a quadratic form: this is the sum for possible couples of objects i, and j of the assignment cost defined as the product of flow by distance.

As such, usually A_{ij}, and B_{ij} are positive real numbers. The matrix B could, in fact, represent a distance matrix (triangular inequality is fulfilled), but this is not necessary. Likewise, the matrix B could be symmetric, but this is not necessary in the general case. Here, we will only consider that the self distance B_{ii} is equal to 0 for all $i \in [n]$. As a consequence, A_{ii} can be considered as equal to 0 for all $i \in [n]$.

2.2 QAP Benchmark

QAP has a lot of applications in real world [9]. As a consequence, a lot of benchmark instances have been proposed to understand problem difficulty, or to design more efficient optimization algorithms according to the features of QAP [22]. The most well-known benchmark of QAP problems is the QAPLib [6]. QAPLib is a collection of instances with real-world problems usually of small size, and larger artificial ones generated with specific properties. The most used artificial instances are probably the two series of Taillard instances (*Taia*, and *Taib*) [32]. In the uniform instances (*Taia*), the distance matrix is a Euclidean distance matrix between random points in a circle, and the flow matrix is random matrix with integer randomly selected between two bounds. The real-like instances (*Taib*) are inspired from some real world problems and mimic some of their properties. The distance matrix is also an Euclidean matrix but where the points are clustered, and the values of the flow matrix are exponentially distributed. Indeed, a lot of entries of the flow matrix are equal to zero. It is well demonstrated that uniform instances are more challenging than real-like instances [34].

Other instances have been proposed. The *Taie* and *Dre* series of instances were specifically designed to be difficult for metaheuristics [11]. Additionally, Stützle *et al.* generated instances which vary two instance parameters systematically; these are related to flow dominance and sparsity [30]. A special case of QAP which is polynomially solveable [20] has been also proposed in order to test "black-box" algorithms. Designing relevant QAP instances and understanding their properties has a lot of interest because doing so can facilitate the testing and improvement of optimization algorithms.

2.3   Features and Problem Difficulty in QAP

Instances of the QAP were first characterised by the notion of flow dominance [35], which measures imbalance in the A and B matrices. A very high dominance value would be obtained if there is a substantial distance between comparatively few locations, or if there is a high degree of flow between only a few facilities. Measurement of sparsity for the two matrices has also been used in the literature [29]: this is the number of zero-entries as a proportion of the n^2.

The QAP has served as one of the main testbeds for fitness landscape analysis of combinatorial spaces and several measures have been shown to be linked to search difficulty in some way. The correlation length — which captures how far apart solutions with related fitness are — and the fitness-distance correlation — the connection between distance and fitness among local optima — were related to the performance of memetic search algorithms for QAP [23]. A set of measurements including some from information theory were computed from walks on QAP landscapes and used to aid in algorithm decisions [26]. Another study focused considered whether landscape measures might be linked to the nature of the instance specification (distance and flow matrices) [33], finding that autocorrelation and the size of plateaus were affected by the number of similar values in the matrices. The local optima space of QAP instances has also been studied [10].

Fourier decomposition has been applied to the QAP: a branch-and-bound approach which operates in the Fourier space [18] has been proposed. Elementary landscape decomposition has also been leveraged [8]; this proved that the QAP, through the prism of the pairwise swap neighbourhood, can be represented as the combination of exactly three elemental landscapes.

3   Definition of QAP-SAT

3.1   Generic QAP-SAT

The idea of QAP-SAT is to define the fitness function as a sum of easy QAP problems, called clauses, which depend only on a few variables. QAP-SAT follows the principle of the MAX-SAT problem, which is defined as the sum of the satisfaction of each clause, where clauses are easy low dimensional pseudo-boolean problems. In QAP-SAT, we say that a clause is satisfied when a candidate solution of the problem reaches the lower bound of the clause. Although there are similarities, a difference to the MAX-SAT space of functions is that the set of functions for QAPs is not a vector space [12]. In general, the sum of two QAP problems defined on $\mathcal{S}_n$ is not a QAP problem on $\mathcal{S}_n$. The QAP space has a bi-linear property: for any matrices A, A', B, B' of dimension $n \times n$, $Q_{A+A',B+B'} = Q_{A,B}+Q_{A,B'}+Q_{A',B}+Q_{A',B'}$. The linear property between QAP problems is preserved when the same distance matrix B is shared. For any distance matrix B, for any positive integer m, and any square matrices $A_1, \ldots A_m$: $Q_{\sum_{\alpha=1}^m A_\alpha,B} = \sum_{\alpha=1}^m Q_{A_\alpha,B}$. In the following, we define a clause for QAP by defining a clause both for flow matrix A, and distance matrix B.

A-*Clause.* A matrix A of dimension $n \times n$ is an A-*Clause.* of size $k > 0$ when $\forall i \in [n]$ $A_{ii} = 0$, and it exists a subset $V_A \subset [n]$ of size k such that: $\forall (i,j) \in V_A^2$, $i \neq j$, $A_{ij} > 0$, and $\forall (i,j) \notin V_A^2$, $A_{ij} = 0$. The left two matrices in Fig. 1 form an example of A-clause.

QAP-SAT Clause. A QAP problem $Q_{A,B}$ is a clause of size k for the matrix B iff A is an A-clause of size k.

When the flow matrix A is an A-clause of size k, the computation of the corresponding clause $Q_{A,B}$ is reduced to the sum of the $k(k-1)$ non-zero terms: $\forall \sigma \in S_n$, $Q_{A,B}(\sigma) = \sum_{(i,j) \in V_A^2, i \neq j} A_{ij} B_{\sigma_i \sigma_j}$ In this case, a lower bound $\mathrm{lb}(Q_{A,B})$ of the $Q_{A,B}$ is: $\ell \sum_{(i,j) \in V_A^2, i \neq j} A_{ij}$ where $\ell = \min\{B_{ij} : (i,j) \in [n]^2\}$. For example for the matrix A_3, this lower bound for $Q_{A_3,B}$ is equal to 10. A B-clause for the distance matrix is defined to ensure that the previous lower bound can be reached for a clause problem $Q_{A,B}$. In this work, without losing generality, the minimum non-zero value of the distance matrix is fixed to 1. This value can be fixed to an arbitrary positive value. In this case, all values would be scaled to the selected non-zero minimum.

B-*Clause.* A matrix B of dimension $n \times n$ is a B-*clause* of size $k > 0$ when $\forall i \in [n]$ $B_{ii} = 0$, and there exists a set $V_B \subset [n]$ of size k such that: $\forall (i,j) \in V_B^2$, $i \neq j$, $B_{ij} = 1$, and $\forall (i,j) \notin V_B^2$, $B_{ij} = M$ where $M > 1$ is the largest possible distance of the problem.

As a consequence, when A_k is an A-clause of size k, and B_k is a B-clause of size k, the minimum of the clause Q_{A_k,B_k} is the lower bound $\mathrm{lb}(Q_{A_k,B_k})$. More generally, a clause $Q_{A,B}$ is said *satisfied* iff the minimum of $Q_{A,B}$ is equal to the lower bound $\mathrm{lb}(Q_{A,B}) = \sum_{(i,j) \in V_A^2, i \neq j} A_{ij}$. The QAP-SAT problem is an aggregation of A-clauses and B-clauses. Let us define the aggregation principle of B-clauses. As for the Hadamard product, let us define the matrix $B \odot B'$ by taking the minimum element by element: $\forall (i,j) \in [n]^2, (B \odot B')_{ij} = \min\{B_{ij}, B'_{ij}\}$. Notice that $\odot$ is an associative operator. We say that a distance matrix B is composed of m_1 B-clauses, with $m_1 > 0$, when it exists m_1 B-clauses $B_1, \ldots, B_{m_1}$, and a matrix C with $\forall i \neq j$, $C_{ij} > 1$, and $\forall i, C_{ii} = 0$, such that $B = B_1 \odot B_2 \odot \ldots \odot B_{m_1} \odot C$.

QAP-SAT. The QAP problem $Q_{A,B}$ is a *QAP-SAT* problem with m A-clauses, and m_1 B-clauses when the matrix B is composed of m_1 B-clauses, and it exists m clauses $Q_{A_1,B}, \ldots, Q_{A_m,B}$ for the same matrix B such that $Q_{A,B}$ is the sum of those m clauses: $Q_{A,B} = \sum_{\alpha=1}^{m} Q_{A_\alpha,B}$. A QAP-SAT problem $Q_{A,B}$ is *satisfied* when all clauses are satisfied, *i.e.* when all clauses reach the lower bound: $\min Q_{A,B} = \sum_{\alpha=1}^{m} \mathrm{lb}(Q_{A_\alpha,B})$.

3.2 Random QAP-k-SAT

When all clauses (A-clauses and B-clauses) have the same size $k > 0$, the QAP-SAT is denoted QAP-k-SAT. The random QAP-k-SAT is designed with the same principle of the classical random k-SAT problem. The k variables of each clause are randomly and independently selected. The random QAP-k-SAT is defined

$$A_3 = \begin{bmatrix} 0&0&0&0&0 \\ 0&0&1&0&2 \\ 0&2&0&0&1 \\ 0&0&0&0&0 \\ 0&3&1&0&0 \end{bmatrix} \quad A^{(3)} = \begin{bmatrix} 0&1&2 \\ 2&0&1 \\ 3&1&0 \end{bmatrix} \quad B_3 = \begin{bmatrix} 0&1&M&M&1 \\ 1&0&M&M&1 \\ M&M&0&M&M \\ M&M&M&0&M \\ 1&1&M&M&0 \end{bmatrix} \quad B^{(3)} = \begin{bmatrix} 0&1&1 \\ 1&0&1 \\ 1&1&0 \end{bmatrix} \quad B = \begin{bmatrix} 0&1&2&4&1 \\ 1&0&4&2&1 \\ 3&8&0&5&3 \\ 3&6&7&0&2 \\ 1&1&2&5&0 \end{bmatrix}$$

Fig. 1. For problem dimension $n = 5$, examples of A-clause and B-clause of size $k = 3$ with $V_A = \{2,3,5\}$, and $V_B = \{1,2,5\}$. $A^{(3)}$, and $B^{(3)}$ are sub-matrix with variables of V_A, and V_B. Distance matrix B composed of $m_1 = 1$ B-clause complementary to matrix B_3.

by 4 basic parameters. The problem size n, the size of the clause k, the number m of A-clauses, and the number m_1 of B-clauses.

For each A-clause, k different variables are randomly selected. In this work, the size of clauses is set to $k = 3$. It is possible to create any random sub-matrix of size k for A-clause. However, in this work for simplicity (same lower bound for example), the A-clauses are based on the same sub-matrix $A^{(3)}$. Only the order of variables is randomly swapped. For each B-clause, k different variables are also randomly selected. Notice that the minimal B-clause is symmetric (see matrix $B^{(3)}$), no need to swap randomly the variables. Several choices can be made to create the complementary matrix C of distances which have non-minimal values. Of course a basic choice would be to create random integer numbers between 2 and a maximal value. However, in this work, we prefer a slightly more sophisticated choice. All values of C are integer values higher or equal than 2. They are selected in order to have geometric distribution of values. Let n_d be the number of values equal to d in the matrix B: $n_d = \sharp\{B_{ij} = d : (i,j) \in [n]^2\}$, and p_d the respective proportion in the matrix: $p_d = n_d/(n(n-1))$. For $d > 1$, the proportion p_d is set in order to have approximately $p_d = p_1^d$ (to the precision of integer values for n_d). Indeed, we set $n_d = \max\{1, \lceil p_1(n(n-1) - \sum_{\delta=1}^{d-1} n_\delta)\rceil\}$. Then, the position of values is randomly distributed in B on available positions in the matrix. As a consequence, the matrix B is not necessary symmetric. Figure 1 is an example of distance matrix. Python code for the generator of QAP-SAT instances, and the instances used in this work are available on the repository: https://gitlab.com/verel/qap-sat.

4   Experimental Analysis

4.1   Experimental Design

In this work, we generate small and medium size instances of QAP-SAT from the dimension $n = 8$ to the dimension $n = 19$. The instance parameters are given in Table 1 to generate a factorial design. 50 instances for each parameter triplet (n, m_1, m) have been generated. Thus, for dimension n lower than 17, $18,000$ instances are generated for problem dimension. To reduce the computation time,

Table 1. Parameter settings of the random QAP-3-SAT instances. The sets corresponds to parameters for dimensions $n = 18, 19$.

Name	Description	Values	
n	Problem dimension	$\{8, 9, \ldots, 17\}$	$\{18, 19\}$
k	Size of clause	3	
m_1	Number of B-clauses (distance matrix)	$\{3, 6, 9, \ldots, 27\}$	$\{3, 9, 15, \ldots, 57\}$
m	Number of A-clauses (flow matrix)	$\{1, 2, 3, \ldots, 40\}$	$\{1, 3, 9, 15, \ldots, 57, 63\}$

for problem dimensions 18 and 19, we reduce the number of instances to 6,000. In total, 192,000 instances have been analyzed. Small size instances can be fully enumerated until dimension $n \leq 13$, then we use a branch and bound algorithm to compute global minima (see next paragraph).

Branch and Bound Algorithm. An exact algorithm is required to find the global minimum of each instance, in particular for medium size instances with dimension larger than 14. First, we test a classical Cplex algorithm with a standard linear transformation of QAP, provided by [27][1], but the computation time is too long for our experimental scenario. For instance, the average computation time for the small dimension $n = 10$, across all m and m_1 values is equal to 50.6 seconds. Indeed, this first experiment – not detailed in this paper for the sake of brevity – is an indication that the QAP-SAT instances can be difficult to solve.

Several Branch and Bound (B&B) algorithms have been proposed to solve QAP from the seminal works based on Gilmore-Lawler bound [16]. In this work, we use a recent and efficient B&B algorithm proposed by Fujii *et al.* [13] for which the MATLAB code is available[2]. The algorithm is based on the Lagrangian doubly non-negative relaxation and Newton-bracketing. Please refer to the original article for details. The goal of this work is not to compare the efficiency of B&B algorithms for solving QAP-SAT instances, but to find the global minimum and estimate the computation time to find it as a possible measure of difficulty.

Robust Taboo Search. We use Taillard's implementation in C of his robust taboo search (ROTS) algorithm for the QAP [31][3]; this is considered a competitive metaheuristic for the domain. The neighbourhood is a random pairwise swap in the permutation, and parameters are kept as those provided in the code: tabu duration is $8n$; aspiration is set at $5n^2$; and runs terminate when the global optimum is found or after 1000 solutions have been visited by the search. The global optimum has been computed for all considered instances; proportional success rate is therefore computed as a metric of performance. We compute the mean for this metric over 30 runs per instance.

---

[1] https://github.com/afcsilva/PMITS-for-QAPVar.
[2] https://sites.google.com/site/masakazukojima1/softwares-developed/newtbracket?pli=1.
[3] http://mistic.heig-vd.ch/taillard/codes.dir/tabou_qap2.c.

4.2   Phase Transition of Satisfaction Probability

A phase transition in combinatorial optimization is characterized by a rapid change between satisfied and non satisfied instances according to a phase parameter. Figure 2 shows the proportion of satisfied instances for which the minimum is the lower bound *i.e.* all clauses of the problem are satisfied with the minimal possible value. As expected, when the number A-clauses m — the number of clauses for the flow matrix — is small, the probability to have an instance satisfied is nearly equal to 1. This probability drops very quickly around a critical value denoted m_c. When the number m of A-clauses is much larger than the number of B-clauses m_1 — the number of clauses for the distance matrix — none of the instances are satisfied. The only exception is for small dimension $n = 8, 9$ for which a large number of B-clauses m_1 gives a full distance matrix of 1, and then all solutions are global optima. In general, this curve seems to describe a sigmoidal shape dropping quickly around a critical value m_c. Indeed, for a given problem dimension, the critical value m_c increases with the number m_1 of B-clauses. For instance, for problem dimension $n = 10$, m_c is around 8 when $m_1 = 9$, and around 15 when $m_1 = 21$. But as problem dimension increases, the variation of m_c according to m_1 is smaller. More details are given in the next Sect. 4.3.

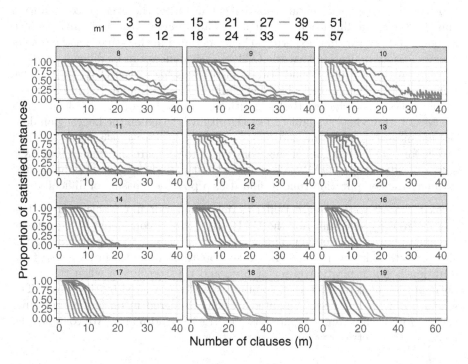

Fig. 2. Proportion of satisfied instances, for which the minimal values is the lower bound, according to the number of A-clauses m. Facet: problem dimension n.

4.3   Estimation of the Phase Transition Parameter

The proportion of satisfied instances (see Fig. 2) seems to follow a logistic function as a function of the number of A-clauses m. If it is true, it is possible to estimate the parameters of the logistic model, and compute the center of symmetry which correspond to inflection point of the logistic model. This center is the critical value of m at the phase transition. First, to estimate the parameters of the logistic regression, we use logit transform: if p follows a logistic model, then $\mathrm{logit}(p) = \log(\frac{p}{1-p})$ follows a linear model, and inversely. If the regression model is $\mathrm{logit}(p) = \beta_0 + \beta_1 m$, the abscissa of the center of the logistic model is $m_c = -\beta_0/\beta_1$.

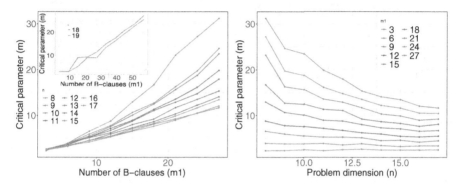

Fig. 3. Critical parameter m_c according to the number of B-clauses m_1 (left), and problem dimension n (right).

The critical value m_c of the logistic model is estimated by the regression of the logit model. Figure 3 shows the critical value m_c of the number of clauses according to the number of minimal B-clauses, and the problem dimension n. Those critical parameters have been estimated using the logit regression. The R^2 values of the regression, which is not presented here to save space, are high. On the 110 possible instance parameters, the average R^2 is 0.9417, and 98 instances have R^2 above 0.9. The worst values of R^2 (minimum is 0.801) are obtained for the smallest problem dimensions where the probability of satisfaction does not reach exactly 0 for large m.

Except for small values of $n = 8$ and 9, the critical value m_c as a function of B-clauses m_1 tends to be linear. Table 2 gives the estimated parameter values for a linear model $m_c = \beta_0 + \beta_1 m_1 + \epsilon$. For n higher than 11, the adjusted R^2 regression quality is high: larger than 0.98. The origin ordinate of the regression is close to 0, and the slope of the regression decreases with problem dimension n. At first, this suggests a linear dependency between critical value m_c and m_1 when n is "large" compared to m_1. Notice that the slope of the linear regression decreases with problem dimension n. From the right side of Fig. 3, for a given value m_1, the critical value m_c seems not always depend linearly on problem dimension n. Only for values small of B-clauses $m_1 \leq 9$, the critical value seems to be independent of problem dimension.

Table 2. Estimated parameter values of the regression model $m_c = \beta_0 + \beta_1 m_1 + \epsilon$.

n	origin β_0	slope β_1	adj. R^2	n	origin β_0	slope β_1	adj. R^2
8	-4.09305	1.258099	0.9702023	14	0.4362908	0.4840685	0.9877075
9	-1.960525	0.9022615	0.9679433	15	0.4972105	0.458738	0.9861831
10	-1.290005	0.8273129	0.9797667	16	0.8578043	0.3977207	0.988515
11	-0.5588647	0.7006147	0.9843299	17	1.104765	0.3881437	0.9974293
12	-0.01016369	0.6367813	0.9872014	18	0.3636075	0.454062	0.9813189
13	0.7395489	0.5280727	0.9923386	19	-0.6242743	0.4427259	0.97458

From the first observations, we can try to explain the critical value m_c as a function of both m_1, and n: close to linear model as a function of m_1, but with a slope decreasing slowly with problem dimension. Moreover, from inspiration of SAT problem for which the phase transition parameter is the ratio between number of clauses and problem dimension, we run the following regression model:

$$m_c = k n^{\alpha_1} m_1^{\alpha_2} + \epsilon$$

where k is a real constant value, and α_1, α_2 are exponents for variables n, and m_1, and ϵ is the noise of the regression model. The parameters of the model can be estimated with a multi-linear regression on the logarithm of m_c: $\log(m_c) = \log(k) + \alpha_1 \log(n) + \alpha_2 \log(m_1)$. We estimate the parameters using the values of $\log(m_c)$ for problem dimension below 17. The adjusted R^2 of this regression is $R^2 = 0.947$ which gives an R^2 coefficient on the value m_c (without log transformation) of $R^2 = 0.898$. The parameters of the regression are: $\alpha_1 = -0.75999$, and $\alpha_2 = 0.90365$, and $\log(k) = 1.65453$. As expected from the previous linear regression, the exponent α_2 for m_1 is close to the value 1, and the exponent α_1 for n is negative between $-1/\sqrt{n}$, and $-1/n$ approximately. As n increases, the slope of the linear relation between m_c, and m_1 decreases. When we test this model on instances with $n \geq 18$, the R^2 is higher, equal to 0.923, which corroborates to the robustness of the model even if more data or a theoretical proof could help to validate it further. So, we hypothesise that $\frac{m}{n^{\alpha_1} m_1^{\alpha_2}}$ is the phase parameter of QAP-3-SAT. To check this hypothesis, Fig. 4 shows the proportion of satisfied instances as a function of this phase parameter. For all problem dimensions, and numbers of B-clauses, the probability to have satisfied instances drops very quickly around the same value k.

4.4 Performance of Optimization Algorithms

Branch and Bound. In the previous section, the Branch and Bound (B&B) algorithm was used to find the global minimum. In this section, it is used to

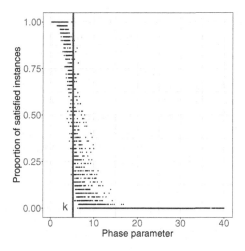

Fig. 4. Probability of satisfied instances according to the phase transition parameter $\frac{m}{n^{\alpha_1}m_1^{\alpha_2}}$ across all instances. $k = \exp(1.65453) \approx 5.23$.

estimate the computational effort to find the global minimum. Notice that the B&B algorithm is used in an optimization scenario to find and to prove global minima, and not in the decision scenario to decide only if a QAP-SAT instance is satisfied. The Fig. 5 shows the average computation time (in seconds) of the B&B algorithm across instances sharing the same parameters. For fair comparison, the algorithm is run on the same computer configuration (Dell HPE DL385 with 512Go RAM, 2 processors AMD-EPYC Milan with 256M cache, and 48 cores at frequency 2.3Ghz each) for all instances. The computation time increases by a larger factor when the problem dimension increases. For a given problem dimension n, the computation time drops from few seconds for instances with low-m, to high value after a threshold value m. Indeed, the computation time seems to follow a sigmoidal shape with the number of clause m. Around a critical value of m, the computation time increases fast.

We analyze the relation between the critical parameter m_c of satisfaction phase transition and the potential critical parameter of B&B computation time. First, we compute a regression using sigmoid function of average computation time: $t(m) = \frac{L}{1+e^{-r(m-m_t)}}$ where L is the maximal value, r the rate of increases, and m_t the inflection of the sigmoid $i.e.$ the critical parameter. Contrary to the previous section, due to noise of the computation time logit transformation can not be used to estimate the parameters. As we know the range of parameters, we use a basic grid search to estimate the regression parameters minimizing the mean square error. Except for $n = 10$ for which the variance of computation time is high and not stable, the R^2 coefficients of the regression are high — over 0.925 — with a median equal to 0.969. This result tends to show that the computation time of B&B follows a sigmoidal shape like in the phase transition. Figure 7 (left) shows the relation between critical parameter m_c of the satisfaction phase transition and the critical parameter of B&B computation time. For a given problem dimension, the relation is nearly linear. The average computation time depends on the phase transition parameter. However, the variance of the computation time critical parameter m_t depends also on the problem dimension. For the same critical parameter m_c, the value of m_t increases with problem dimension, and the range of variation with m_c gets smaller as problem dimension increases. Indeed, we notice that the average of the maximum time L across m_1 value for a given problem dimension n is approximately given by $\gamma(2.043 + 0.476(n-8))$. To prove that a candidate solution is a global minimum, B&B needs to cross a large part of the search anyway.

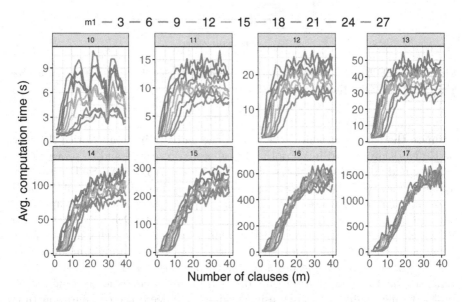

Fig. 5. Average computation time of B&B algorithm to find the global minimum. Problem dimensions $n = 10$–17.

Robust Taboo Search. Figure 6 presents the success rate of ROTS for all considered instances. There is a very high negative correlation between success rate, and number of evaluations to reach global minimum ($\rho = -0.9999$). So only the success rate is analyzed in the following. In the Figure, instances are split into facets according to problem dimension, n, and split by m_1 as individual lines. The horizontal axis is number of clauses in the instance, m. It follows that a single line in one of the plots represents, for all instances of the specified size and m_1, the mean ROTS success rate.

Notice from Fig. 6 that the success rate decreases with increasing m for all problem dimensions. This effect is much stronger for larger problem sizes, however. From surveying individual facets it can be observed that lower values of m_1 are associated with lower success rates, although the disparities between low-m_1

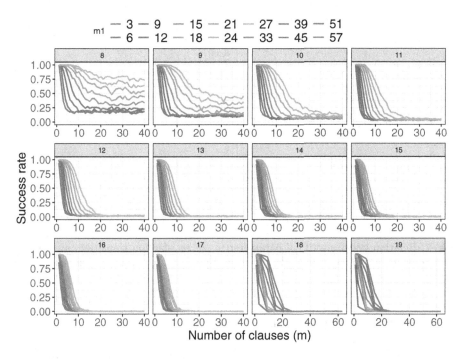

Fig. 6. Success rate with varying number of clauses, M, for Robust Taboo Search. Problem dimensions $n = 8$–19.

and high-m_1 instances becomes substantially less for higher problem dimensions (compare, for example, the facet for problem size 8 with that of size 16). For low values of m_1, there is a steep decrease in success rate at approximately $m = 5$. For larger m_1, the drop in success rate happens at a larger m, and for smaller problem sizes there is not a dramatic vertical drop for them. For problem sizes at 14 or greater, however, instances with all values of m_1 experience a steep drop in success rate between m being 5 and approximately 21.

We also analyze the relation between the critical parameter m_c of satisfaction phase transition, and the success rate of tabu search. Like for B&B we obtain the estimate of a sigmoid model for success rate as a function of the number of clauses m using a basic grid search minimizing mean square error. Except for very small values of B-clauses m_1, R^2 coefficients are high — always over 0.9436 with median equals to 0.9972. In contrast to B&B, the correlation with the critical parameter of satisfaction phase transition is very high: $\rho = 0.9904$. The R^2 of the linear regression is $R^2 = 0.9811$. The slope of the linear relation is 0.542. This result shows that the difficulty to solve a QAP-SAT instance for tabu search is highly dependent on the position of the instance against the phase transition parameter. Recall that the tabu search stops when the global minimum is found; maybe the B&B needs some additional time (which could depend on the problem dimension) to prove that the solution is the global minimum.

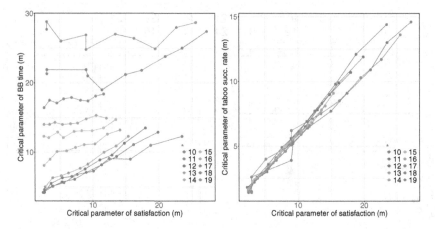

Fig. 7. Relation between critical parameter m_c of satisfaction transition, and critical parameter m_t for B&B average computation time (left), for taboo search success rate (right)

5   Discussion and Future Work

In this work, QAP-SAT instances are designed using a submodularity principle by defining low-dimensional easy problems called "clauses" to solve jointly like in SAT problems. Each flow and distance matrix is composed of basic sub-matrices, and the agreement between sub-matrices in the flow and distance matrices sharply tunes the difficulty of the instance. Although limited to medium size instances with problem dimension up to 19, the large experimental analysis shows that the QAP-SAT problem shows a phase transition according to number of clauses in the matrices and the problem dimension. Supported by those first experiments, we suggest an order phase parameter that explains the phase transition. Moreover, the problem performance of the Branch & Bound algorithm and robust taboo search are correlated with phase transition.

This initial work raises many open questions. First, it is now possible to test optimization algorithms using QAP-SAT instances as a benchmark, including large size instances, and compare QAP-SAT against existing QAP benchmarks to stress the difference between them. Moreover, the modular design of QAP-SAT with clauses motivates the decomposition of existing real-world QAP instances into easy sub-problems. Additionally, it could inspire the definition of new properties and metrics on each matrix, and also jointly between matrices. More broadly, it would be relevant to analyse the fitness landscape to understand the phase transition and problem difficulty in QAP-SAT. From the theoretical side, a proof using existing mathematical techniques is required to support the proposed phase transition parameter based on experiments. In this work, sub-problems of dimension 3 (QAP-3-SAT) have been studied using a specific shape for A-clauses, and B-clauses, but future works could extend the analysis to a broader class of clauses, or to relax the satisfiability condition of a clause, and the impact on phase transition parameters. Such extended analysis is likely to be a new field of questions for QAP-related structural properties and problem difficulty.

References

1. Achary, T., Pillay, S., Pillai, S.M., Mqadi, M., Genders, E., Ezugwu, A.E.: A performance study of meta-heuristic approaches for quadratic assignment problem. Concurr. Comput.: Pract. Experience **33**(17), e6321 (2021)
2. Ahuja, R.K., Orlin, J.B., Tiwari, A.: A greedy genetic algorithm for the quadratic assignment problem. Comput. Oper. Res. **27**(10), 917–934 (2000)
3. Benlic, U., Hao, J.K.: Memetic search for the quadratic assignment problem. Expert Syst. Appl. **42**(1), 584–595 (2015)
4. Biroli, G., Cocco, S., Monasson, R.: Phase transitions and complexity in computer science: an overview of the statistical physics approach to the random satisfiability problem. Phys. A **306**, 381–394 (2002)
5. Buhmann, J.M., Dumazert, J., Gronskiy, A., Szpankowski, W.: Phase transitions in parameter rich optimization problems. In: 2017 Proceedings of the Fourteenth Workshop on Analytic Algorithmics and Combinatorics (ANALCO), pp. 148–155. SIAM (2017)
6. Burkard, R.E., Karisch, S.E., Rendl, F.: QAPLIB-a quadratic assignment problem library. J. Global Optim. **10**, 391–403 (1997)
7. Cheeseman, P.C., Kanefsky, B., Taylor, W.M.: Where the really hard problems are. In: IJCAI, vol. 91, pp. 331–337 (1991)
8. Chicano, F., Luque, G., Alba, E.: Elementary landscape decomposition of the quadratic assignment problem. In: Proceedings of the 12th annual conference on Genetic and evolutionary computation, pp. 1425–1432 (2010)
9. Commander, C.W.: A survey of the quadratic assignment problem, with applications (2005)
10. Daolio, F., Verel, S., Ochoa, G., Tomassini, M.: Local optima networks of the quadratic assignment problem. In: IEEE Congress on Evolutionary Computation, pp. 1–8. IEEE (2010)
11. Drezner, Z., Hahn, P.M., Taillard, É.D.: Recent advances for the quadratic assignment problem with special emphasis on instances that are difficult for meta-heuristic methods. Ann. Oper. Res. **139**, 65–94 (2005)
12. Elorza, A., Hernando, L., Lozano, J.A.: Characterizing permutation-based combinatorial optimization problems in fourier space. Evol. Comput., 1–39 (2022)
13. Fujii, K., Ito, N., Kim, S., Kojima, M., Shinano, Y., Toh, K.C.: Solving challenging large scale QAPs. arXiv preprint: arXiv:2101.09629 (2021)
14. Gent, I.P., Walsh, T.: The sat phase transition. In: ECAI, vol. 94, pp. 105–109. PITMAN (1994)
15. Gent, I.P., Walsh, T.: The TSP phase transition. Artif. Intell. **88**(1–2), 349–358 (1996)
16. Gilmore, P.C.: Optimal and suboptimal algorithms for the quadratic assignment problem. J. Soc. Ind. Appl. Math. **10**(2), 305–313 (1962)
17. Hartmann, A.K., Weigt, M.: Phase Transitions in Combinatorial Optimization Problems: Basics, Algorithms and Statistical Mechanics. John Wiley & Sons, Hoboken (2006)

18. Kondor, R.: A fourier space algorithm for solving quadratic assignment problems. In: Proceedings of the Twenty-first Annual ACM-SIAM Symposium on Discrete Algorithms, pp. 1017–1028. SIAM (2010)

19. Koopmans, T.C., Beckmann, M.: Assignment problems and the location of economic activities. Econometrica: J. Econometric Soc., 53–76 (1957)

20. Laurent, M., Seminaroti, M.: The quadratic assignment problem is easy for Robinsonian matrices with Toeplitz structure. Oper. Res. Lett. **43**(1), 103–109 (2015)

21. Leyton-Brown, K., Hoos, H.H., Hutter, F., Xu, L.: Understanding the empirical hardness of NP-complete problems. Commun. ACM **57**(5), 98–107 (2014)

22. Loiola, E.M., De Abreu, N.M.M., Boaventura-Netto, P.O., Hahn, P., Querido, T.: A survey for the quadratic assignment problem. Eur. J. Oper. Res. **176**(2), 657–690 (2007)

23. Merz, P., Freisleben, B.: Fitness landscape analysis and memetic algorithms for the quadratic assignment problem. IEEE Trans. Evol. Comput. **4**(4), 337–352 (2000)

24. Monasson, R.: Introduction to phase transitions in random optimization problems. arXiv preprint: arXiv:0704.2536 (2007)

25. Monasson, R., Zecchina, R., Kirkpatrick, S., Selman, B., Troyansky, L.: Determining computational complexity from characteristic? Phase transitions? Nature **400**(6740), 133–137 (1999)

26. Pitzer, E., Beham, A., Affenzeller, M.: Automatic algorithm selection for the quadratic assignment problem using fitness landscape analysis. In: Middendorf, M., Blum, C. (eds.) Evolutionary Computation in Combinatorial Optimization. Lecture Notes in Computer Science, vol. 7832, pp. 109–120. Springer, Berlin (2013). https://doi.org/10.1007/978-3-642-37198-1_10

27. Silva, A., Coelho, L.C., Darvish, M.: Quadratic assignment problem variants: a survey and an effective parallel memetic iterated TABU search. Eur. J. Oper. Res. **292**(3), 1066–1084 (2021)

28. Smith-Miles, K., Lopes, L.: Measuring instance difficulty for combinatorial optimization problems. Comput. Oper. Res. **39**(5), 875–889 (2012)

29. Smith-Miles, K.A.: Towards insightful algorithm selection for optimisation using meta-learning concepts. In: 2008 IEEE International Joint Conference on Neural Networks (IEEE World Congress on Computational Intelligence), pp. 4118–4124. IEEE (2008)

30. Stützle, T., Fernandes, S.: New benchmark instances for the QAP and the experimental analysis of algorithms. In: Gottlieb, J., Raidl, G.R. (eds.) Evolutionary Computation in Combinatorial Optimization. Lecture Notes in Computer Science, vol. 3004, pp. 199–209. Springer, Berlin (2004). https://doi.org/10.1007/978-3-540-24652-7_20

31. Taillard, É.: Robust taboo search for the quadratic assignment problem. Parallel Comput. **17**(4–5), 443–455 (1991)

32. Taillard, E.D.: Comparison of iterative searches for the quadratic assignment problem. Locat. Sci. **3**(2), 87–105 (1995)

33. Tayarani-N, M.H., Prügel-Bennett, A.: Quadratic assignment problem: a landscape analysis. Evol. Intel. **8**, 165–184 (2015)

34. Verel, S., Daolio, F., Ochoa, G., Tomassini, M.: Sampling local optima networks of large combinatorial search spaces: the QAP case. In: Auger, A., Fonseca, C., Lourenc, N., Machado, P., Paquete, L., Whitley, D. (eds.) Parallel Problem Solving from Nature - PPSN XV. Lecture Notes in Computer Science(), vol. 11102, pp. 257–268. Springer, Cham (2018). https://doi.org/10.1007/978-3-319-99259-4_21

35. Vollmann, T.E., Buffa, E.S.: The facilities layout problem in perspective. Manage. Sci. **12**(10), B–450 (1966)

36. Weigt, M., Hartmann, A.K.: Number of guards needed by a museum: a phase transition in vertex covering of random graphs. Phys. Rev. Lett. **84**(26), 6118 (2000)
37. Yadav, N., Murawski, C., Sardina, S., Bossaerts, P.: Phase transition in the knapsack problem. arXiv preprint: arXiv:1806.10244 (2018)

Hardest Monotone Functions
for Evolutionary Algorithms

Marc Kaufmann$^{(\boxtimes)}$, Maxime Larcher, Johannes Lengler, and Oliver Sieberling

Department of Computer Science, ETH Zürich, Zürich, Switzerland
`marc.kaufmann@inf.ethz.ch`

Abstract. The hardness of optimizing monotone functions using the
$(1 + 1)$-EA has been an open problem for a long time. By introducing
a more pessimistic stochastic process, the partially-ordered evolutionary
algorithm (PO-EA) model, Jansen proved a runtime bound of $O(n^{3/2})$.
In 2019, Lengler, Martinsson and Steger improved this upper bound to
$O(n \log^2 n)$ leveraging an entropy compression argument. We continue
this line of research by analyzing monotone functions that may vary
at each step, so-called dynamic monotone functions. We introduce the
function Switching Dynamic BinVal (SDBV) and prove, using a combinatorial argument, that for the $(1 + 1)$-EA, SDBV is drift minimizing
within the class of dynamic monotone functions. We further show that
the $(1 + 1)$-EA optimizes SDBV in $\Theta(n^{3/2})$ generations. Therefore, our
construction provides the first explicit example which realizes the pessimism of the PO-EA model. Our simulations demonstrate matching
runtimes for both static and self-adjusting $(1, \lambda)$ and $(1 + \lambda)$-EA. We
additionally demonstrate, devising an example of fixed dimension, that
drift minimization does not equal maximal runtime.

Keywords: hardest functions · fitness landscape · precise runtime
analysis · drift minimization · $(1 + 1)$-EA · Switching Dynamic Binary
Value · dynamic environments · evolutionary algorithm

1 Introduction

Drift analysis has become a staple in the analysis of evolutionary algorithms
and more generally of stochastic processes. Roughly speaking, the idea is to
consider a *potential* function which captures the advancement of the process and
to understand how this quantity varies *in expectation* at each step. The more this
quantity *drifts towards* that optimum the faster we expect the optimization to
be. Under which condition one may apply drift theorems, and what bounds they
provide is now well understood, and we refer to Lengler [27] for an introduction
to known results and their applications.

How can we extend a runtime analysis so that it covers not just a single
fitness function, but instead a whole class of fitness landscapes? In [18], Jansen

M. Kaufmann—The author was supported by the Swiss National Science Foundation
[grant number 200021_192079].

ⓒ The Author(s), under exclusive license to Springer Nature Switzerland AG 2024
T. Stützle and M. Wagner (Eds.): EvoCOP 2024, LNCS 14632, pp. 146–161, 2024.
https://doi.org/10.1007/978-3-031-57712-3_10

introduced a random process which he called the Partially-Ordered Evolutionary Algorithm (PO-EA)—named after its key component, a partial ordering on the search space $\{0, 1\}^n$—which can be seen as a *pessimistic abstraction* of the (1+1)-EA operating on linear functions with positive weights. It "shares" with those functions (as well as with the larger class of dynamic monotone functions) the unique global optimum, in the sense that the string consisting of only 1 s is the unique search point which the PO-EA will not replace by any offspring. However, it does not operate on a fitness landscape defined by any fitness function.

One potential drawback of this model is that it may be overly pessimistic compared to the most difficult fitness landscapes which the $(1 + 1)$-EA may encounter. To see why, let us first explain informally how the PO-EA operates.

We say that a search point $x \in \{0, 1\}^n$ *dominates* another search point $y \in \{0, 1\}^n$ if it dominates y bit-wise, that is, there is no position i such that $y_i = 1$ but $x_i = 0$. Many search points are not comparable in this way—for instance with $n = 3$, $x = 101$ and $y = 010$ are not comparable—so this indeed only defines a partial order on the search space. Still, the string consisting of all 1 s is comparable to and dominates any other search point, which motivates the choice of this partial order as a measure of optimization progress. The PO-EA works as follows: in any situation where the parent and the offspring are *comparable*[1], the parent for the subsequent generation is chosen as the one that dominates the other—note that in this case, the selection indeed mimics that of the optimization by the $(1 + 1)$-EA of any monotone function. When the parent and offspring are *incomparable*, then the PO-EA chooses as the next parent the one with fewer 1-bits.

In other words, in this second case, the algorithm deliberately stays or moves away from the optimum in terms of Hamming distance and thus gives a pessimistic estimate of optimization progress. The PO-EA model was subsequently extended by Colin, Doerr and Férey to enable parametrizing the degree of pessimism [2]. In their model, with some probability q the random process takes the pessimistic route and opts for standard selection, that is, picking the survivor based on the Hamming distance to the optimum otherwise.

The PO-EA model is useful because it gives upper bounds for the runtime of the (1+1)-EA on some of the most intensively studied classes of functions: *linear functions* [37] (including the popular benchmarks ONEMAX and BINVAL) and *monotone functions* [7, 26] (including the class of HOTTOPIC functions). It also applies to recent generalizations of these classes to dynamic environments [23, 30, 31]. In particular, the results by Jansen [18] and Colin, Doerr and Férey [2] imply that the $(1 + 1)$-EA with standard bit mutation finds the optimum in all of these cases in time $O(n \log n)$ if the mutation rate is χ/n for any constant $\chi < 1$, and in time $O(n^{3/2})$ if[2] $\chi = 1$. However, for $\chi > 1$ it was shown both for linear and for static monotone functions that those are strictly easier than the PO-EA model. In particular, for $\chi = 1$ the $(1 + 1)$-EA can optimize linear functions in $O(n \log n)$ steps [37] and static monotone functions in $O(n \log^2 n)$

---

[1] That is, either the offspring dominates the parent, or vice-versa.
[2] The latter result was claimed in [18], but the proof contained an error. It was later proven in [2].

steps [28], while the PO-EA needs $\Theta(n^{3/2})$ steps [2]. Moreover, the $(1 + 1)$-EA can solve both linear and (static) monotone functions efficiently for some values $\chi > 1$ for which the PO-EA gives exponential upper bounds. This leads to the following question.

Question 1 (Colin, Doerr, Férey [2]). Is the PO-EA a too pessimistic model for any monotonic function or is there a not-yet discovered monotonic function with optimization time $\Theta(n^{3/2})$?

While the above question was phrased with static monotone functions in mind, the aforementioned results in [28] show that only *dynamic* monotone functions might possibly achieve this bound. Answering this question would allow to "extend" the definition of the PO-EA (or at least its *concept*) to other optimization algorithms that keep a population with more than one element, or that generate more than one child at each step. Knowing the *hardest* monotone functions in terms of expected optimization time or at least a hard function for the $(1 + 1)$-EA would provide a natural 'hard' candidate for other algorithms, and notably its extensions such as the $(\mu + 1)$-EA, the $(1 + \lambda)$-EA or the $(1, \lambda)$-EA. In the context of *self-adjusting* optimization, Kaufmann, Larcher, Lengler and Zou [23] conjectured that ADVERSARIAL DYNAMIC BINVAL (ADBV for short, see Sect. 2 for a formal definition) is the hardest (dynamic) monotone function to optimize. While their question was only formulated for the SA-$(1, \lambda)$-EA, it naturally extends to other optimization algorithms.

Question 2 (Kaufmann, Larcher, Lengler, Zou [23]). What is the hardest monotone function to optimize for the $(1 + 1)$-EA and for other optimization algorithms?

In this paper we will address both these questions and show that there exist dynamic monotone functions which are asymptotically as hard as the PO-EA model.

Regarding the second question, we show that the conjectured ADBV is *not* the hardest function to optimize for the $(1 + 1)$-EA in two aspects.

Firstly, we show that the ADBV construction is not optimal when the distance from the optimum is larger than $n/2$, and we instead propose a modification which we call Switching Dynamic BinVal (SDBV). For this modified candidate, we then prove that among all dynamic monotone functions, SDBV has the minimal drift (expected progress towards the optimum), and that it does indeed have a runtime of $\Theta(n^{3/2})$ for $\chi = 1$, as the PO-EA. This gives an answer to Question 1. Nevertheless, we also show that SDBV is in general *not* the hardest dynamic monotone function *in terms of runtime*. Indeed, we show by numerical calculations that it is not the hardest function for any odd $n \in [9, 45]$. Hence, we give a *negative* answer to the following, plausible question.

Question 3. In the class of dynamic monotone functions, is the function which minimizes the drift towards the optimum also the function which leads to the largest runtime?

Our negative answer to Question 3 can be seen as complementary to the work of Buskulic and Doerr [1], who asked a similar question for *algorithms* instead of *landscapes*. More precisely, they showed that for the $(1+1)$-EA on ONEMAX, for most fitness levels between $n/2$ and $2n/3$ the optimal mutation strengths are higher than those which maximize drift. This disproved a conjecture made in [13]. Hence, the optimal algorithm is more risk-affine than drift-maximizing one. The result in [1] was driven by the desire for *precise runtime analysis* beyond the leading order asymptotics, in particular the result by Doerr, Doerr and Yang that the unary unbiased blackbox complexity of ONEMAX is $n \ln n - cn + / - o(n)$ for a constant $0.2539 \leq c \leq 0.2665$ [4]. As [1] show that the best *algorithm* is not the drift-maximizing one, we show that the hardest *fitness landscape* is not the drift minimizing one.

Summary of Our Contribution. The following summarizes our contribution[3]:

- We provide an explicit construction of a function called SWITCHING DYNAMIC BINVAL (SDBV) and show that it minimizes the drift in the number of zeros at every point; we also provide a runtime analysis. This is the first construction of an explicit function which matches the bounds derived by the PO-EA framework with parametrized pessimism. In particular, SDBV is the first dynamic monotone function which cannot be optimized by the considered class of algorithms in $\tilde{O}(n)$ generations[4].
- We give a proof of its drift minimization - in its function class - using exclusively combinatorial arguments
- We show that drift minimization does not imply maximization of expected runtime, thus extending our understanding of hardness of fitness landscapes.
- We complement our findings with simulations which match our proved bounds, extending experimentally also the results to a larger class of algorithms: $(1 + 1)$-EA, $(1 + \lambda)$-EA, $(1, \lambda)$-EA, SA-$(1, \lambda)$-EA. This provides evidence towards a wider conjecture of Kaufmann, Larcher, Lengler and Zou in [23].

We begin by stating our first main result, which holds for any mutation rate.

Theorem 4. *Switching Dynamic Binary Value is the dynamic monotone function which minimizes the drift in the number of zeros at every search point for the $(1+1)$-EA with any mutation rate.*

Next, we show that, perhaps surprisingly, the minimization of drift does not equate to maximal expected runtime:

Theorem 5. *The drift minimizing function for the $(1+1)$-EA is not the function with the largest expected optimization time.*

The computation of the expected optimization time of Switching Dynamic Binary Value is our final key result.

---

[3] We note that due to page limitations, we omit many of the proofs. The full version is provided in the supplementary material.

[4] Here the $\tilde{O}(n)$ hides potential polylogarithmic factors of n.

Theorem 6. *The* $(1+1)$*-EA with standard mutation rate* $1/n$ *optimizes Switch-ing Dynamic Binary Value in expected* $\Theta(n^{3/2})$ *generations.*

1.1   Related Work

Easiest and Hardest Functions. In [8], Doerr, Johannsen and Winzen showed that ONEMAX is the easiest among all functions with unique global optimum on $\{0,1\}^n$ for the $(1+1)$-EA, meaning that it minimizes the runtime. This has been extended to other algorithms [36] and has been strengthened into stochastic domination formulations in various situations [3,20,37]. On the other hand, it has recently been shown that ONEMAX is not the easiest function with respect to the probability of increasing the fitness, which is important for self-adapting algorithms that rely on this probability [22]. While the hardest among *all* pseudo-Boolean functions is also known [14], it is a long-standing open problem to find the hardest *linear* function. A long-standing folklore conjecture is that this is the BINVAL function, but as of today, this conjecture remains unresolved.

Linear Functions. The seminal paper of Witt [37] showed that all linear functions are optimized in time $O(n \log n)$ by the $(1+1)$-EA with standard bit mutation of rate χ/n, for any constant χ. Recently, Doerr, Janett and Lengler have generalized this to arbitrary unbiased mutation operators as long as the probability of flipping exactly one bit is $\Theta(1)$, and the expected number of flipped bits is constant [12]. It remains open whether this result extends to so-called *fast mutation operators* [10] if they have unbounded expectation. Nevertheless, it is well understood that this class of functions is considerably easier than the PO-EA framework.

Monotone Functions. Monotone functions are a natural generalization of linear functions and one of the few general classes of fitness landscapes that have been thoroughly theoretically investigated. The first result was by Doerr, Jansen, Sudholt, Winzen and Zarges [7], who showed that for a mutation rate of χ/n with $\chi \geq 13$, the $(1+1)$-EA needs exponential time to find the optimum. The constant 13 was later improved to 2.13.. [32], and the result was generalized to a large class of other algorithms by Lengler [26]. He invented a special type of monotone functions called HOTTOPIC functions, which proved to be a hard instance for many evolutionary algorithms if the mutation rate is large. Moreover, Lengler and Zou showed that HOTTOPIC instances are also hard for arbitrarily *small* constants $\chi > 0$ for the $(\mu+1)$-EA if it is used with a large constant population size [33]. As discussed earlier, all upper bounds from the PO-EA framework apply for the $(1+1)$-EA on all monotone functions, including the $O(n^{3/2})$ upper bound for $\chi = 1$. However, Lengler, Martinsson and Steger could strengthen this bound to $O(n \log^2 n)$, showing that (static) monotone functions are strictly easier than the PO-EA framework. Moreover, they showed that the bound $O(n \log^2 n)$ also holds for $\chi = 1 + \varepsilon$ for a sufficiently small constant $\varepsilon > 0$. In this regime, the PO-EA only gives exponential upper bounds.

Dynamic Monotone Functions. Dynamic Monotone Functions were first considered by Lengler and Schaller [31], who studied linear functions with posi-tive weights, where the weights were re-drawn in each generation. It was later argued that the Dynamic Binary Value Function DYNAMIC BINVAL could be obtained as a limit from dynamic linear functions if the weights are maximally skewed [29,30]. Those instances can also be hard to optimize for the $(1+1)$-EA if the mutation rate is large. This is even true if the dynamic environment switches always between the same two functions [17]. Dynamic Monotone Functions were then used in [23] to provide instances on which a self-adapting $(1,\lambda)$-EA may fail. Due to those hardness results, dynamic functions are natural candidates for hard functions within the PO-EA framework. Indeed, [23] also introduced the ADBV construction and conjectured that it is hardest among all dynamic monotone functions. We disprove this conjecture in this paper, but prove that a variant, SDBV, is drift minimizing, though not hardest.

1.2 Organization

The remainder of the paper is organized as follows: we begin with some basic definitions and the description of algorithms as well as the benchmark functions in Sect. 2. In Sect. 3 we construct the function SWITCHING DYNAMIC BINVAL and show that it minimizes Hamming drift in the class of dynamic monotone functions. Due to space constraints, we are unable to provide a complete proof, and instead refer to the full version. Section 4 follows with an explicit computation which shows that despite minimizing the drift, SDBV does not maximize the expected runtime in its function class. In Sect. 5, we show matching upper and lower runtime bounds; again, the details are omitted for brevity, but may be found in the full version. Section 6 contains our experiments, followed by the Conclusion.

2 Preliminary

Throughout this paper, n denotes an arbitrary positive integer, and we consider a stochastic process on the search space $\{0,1\}^n$. Given $x \in \{0,1\}^n$, we let $Z(x) = n - \sum_{i=1}^{n} x_i$ as the number of zero-bits of x. For $x,y \in \{0,1\}^n$, we say that x *dominates* y—which we write $x \geq y$—if $x_i \geq y_i$ for all $1 \leq i \leq n$. A function is called *monotone* if $f(x) > f(y)$ whenever x dominates y, with strict inequality in at least one of the components. A sequence of functions $(f_t)_{t\geq 0}$ is said to be *dynamic monotone* if f_t is monotone for each t. We abuse the terminology slightly and talk of *a dynamic monotone function* rather than a dynamic monotone sequence of functions.

Algorithms. The theoretical analysis presented in this paper focuses on the classical $(1+1)$-EA with standard bit mutation and mutation rate $1/n$, as follows. At each step, the parent x generates a single child y by flipping every bit with probability $1/n$. If the fitness improves, i.e. if $f(y) > f(x)$, then y is selected as the new parent for the next generation. When the fitness stagnates, i.e. if $f(y) = f(x)$, we choose the next parent randomly among x, y. Otherwise y is discarded and x remains the parent for the next generation. The pseudocode for the $(1+1)$-EA may be found in Algorithm 1.

Algorithm 1: $(1+1)$-EA with mutation rate $1/n$, initial start point x^{init} for maximising a fitness function $f\colon \{0,1\}^n \to \mathbb{R}$.

Initialization: Set $x^0 = x^{\text{init}}$;
Optimization: for $t = 0, 1, \dots$ do
> **Mutation**: $y^t \leftarrow$ mutate x^t by flipping each bit indep. with probability $1/n$.
> **Selection**: $x^{t+1} \leftarrow \arg\max\{f(x^t), f(y^t)\}$, breaking ties randomly.

While our proofs are restricted to the $(1+1)$-EA, our simulations described in Sect. 6 confirm our theoretical findings for a larger class of algorithms. Namely we provide simulations for the $(1+\lambda)$-EA, the $(1,\lambda)$-EA, as well as their self-adjusting versions. We refer the interested reader to [9,16,19,24,34] and [11,16, 26,34,35] respectively for the exact pseudocode and known results for the static versions and to [5,15,21–23,25] for the self-adjusting variants.

Benchmarks. Monotone functions serve as a classical benchmark to understand how evolutionary algorithms behave and how fast optimization happens. This choice is motivated by the large size of this function class, for which theoretical analysis is 'relatively easy' (or at least possible). Despite their seeming simplicity (flipping a 0-bit to 1 always increases the fitness), the optimization is non-trivial, as was shown in a series of papers [6,7,26,28,33]. Two notable members are ONEMAX, which simply counts the number of bits set to 1 in x, and BINVAL which returns the value of x read in base 2.

$$\text{OM}(x) = \text{ONEMAX}(x) = \sum_{i=1}^{n} x_i, \quad \text{and} \quad \text{BINVAL}(x) = \sum_{i=1}^{n} 2^i x_i,$$

In this paper we focus on a few *dynamic* monotone functions for which we show optimization is hard.

Here, we consider two dynamic adaptations of BINVAL which we call ADVERSARIAL DYNAMIC BINVAL and FRIENDLY DYNAMIC BINVAL and which we respectively abbreviate to ADBV and FDBV. Given some reference point $x^t \in \{0,1\}^n$, those are respectively defined as

$$\text{ADBV}_{x^t}(x) = \sum_{i=1}^{n} 2^i x_{\sigma_A(i)} \quad \text{and} \quad \text{FDBV}_{x^t}(x) = \sum_{i=1}^{n} 2^i x_{\sigma_F(i)},$$

where $\sigma_A = \sigma_{A,x^t}$ and $\sigma_F = \sigma_{F,x^t}$ are permutations of $[1,n]$ that depend on x^t and that satisfy[5] $\sigma_A(i) \geq \sigma_A(j)$ whenever $x_i^t \leq x_j^t$ and $\sigma_F(i) \leq \sigma_F(j)$ whenever

---

[5] Note that there are several ways of choosing σ_A, σ_F satisfying those properties, as we may permute any 0-bit with another 0-bit, and similarly for the 1-bits. As properties we study hold regardless of the precise selection of σ_A, σ_F, we make a slight abuse of notation and talk of *the* function ADBV and *the* function FDBV.

$x_i^t \leq x_j^t$. In other words, ADBV orders the bits in such a way that the binary value of x^t is *minimized*, while FDBV orders them so that the value of x^t is *maximized*. Note that those definitions ensure that (and this is the property we are interested in), in the selection phase of the $(1+1)$-EA, every non-dominated offspring is accepted when optimizing ADBV; on the contrary, when optimizing FDBV, we only accept offspring which dominates the parent. The name ADBV was coined by Kaufmann, Larcher, Lenger and Zou [21,23] because this choice of dynamic function makes it *deceptively* easy to find a fitness improvement: as each 1-bit of x^t has a lower weight than all 0-bits, a mutation that improves the fitness as soon as a single 0-bit flips to 1, meaning that there is a decently high chance of moving away from the optimum even though the fitness increases. This is for instance not the case with ONEMAX, where two 1-bit flips outweigh a single 0-bit flip and for which a step that increases the fitness necessarily reduces the distance to the optimum. Here we choose to name the second function FDBV since the permutation is selected to satisfy the opposite property of ADBV.

Our results are focused on a last dynamic monotone function which combines both functions above. More precisely, we define SWITCHING DYNAMIC BINVAL, abbreviated to SDBV, as follows

$$\text{SDBV}_{x^t}(x) = \begin{cases} \text{ADBV}_{x^t}(x) & \text{if } \text{OM}(x) \geq n/2; \\ \text{FDBV}_{x^t}(x) & \text{if } \text{OM}(x) < n/2. \end{cases}$$

In this paper, whenever we run any of the evolutionary algorithms introduced in the previous section on ADBV, FDBV or SDBV, we always choose the reference point for those functions as the parent x^t at time t. For this reason, we sometimes omit the subscript x_t and simply write $\text{ADBV}(x), \text{FDBV}(x), \text{SDBV}(x)$ when the parent is clear from context.

Further Tools. Our proofs use classical bounds and results from probability theory. Namely we use the Law of Total Variance, Kolmogorov's Inequality as well as Chernoff's inequality. Due to space constraints we are not able to provide the proofs in full detail, and we choose to omit the precise theorem statements.

3   SDBV Minimizes the Drift Towards the Optimum

The goal of this section is to give a sketch of the key steps in the proof of Theorem 4 and show that SDBV minimizes the drift at every point among all dynamic monotone functions. Due to space constraints we are unable to provide the comprehensive argument; instead we only provide an intuition of the proof by describing the main steps, and we refer to the full version for the complete reasoning. Our proof relies on the observation that, when optimising some function f by the $(1+1)$-EA, the drift of $Z(x^t)$ does not depend on the precise value we assign to each point, but only on which search points are accepted when compared to the parent. In particular, this observation allows us to derive a more 'local' sufficient condition for minimising the drift as expressed in the lemma below.

Lemma 7. *Let $x \in \{0,1\}^n$ be an arbitrary search point. For $i, j \geq 0$ define $A_{i,j}$ as the set of all points that may be reached from x by flipping exactly i zero-bits to one, and j one-bits to zero. Assume that m is a monotone function such that for every monotone f, and every $i > j > 0$ the following quantity*

$$\sum_{y \in A_{i,j}} \left(\mathbf{1}_{f(y)>f(x)} - \mathbf{1}_{m(y)>m(x)} \right) - \sum_{y \in A_{j,i}} \left(\mathbf{1}_{f(y)>f(x)} - \mathbf{1}_{m(y)>m(x)} \right), \quad (1)$$

is non-negative. Then m minimizes the drift at x.

Note that Theorem 4 follows from the above statement if we can prove that for all $x \in \{0,1\}^n$ the quantity (1) is non-negative when replacing m by SDBV_x. Since we always consider (dynamic) monotone functions, accepting or rejecting some search points forces us to also accept or reject all other points which dominate or are dominated by the former. In particular, rejecting elements from $A_{i,j}$ forces us to reject those in $A_{j,i}$ they dominate, and vice versa, accepting some elements from $A_{j,i}$ forces us to accept those in $A_{i,j}$ that dominate them. The second lemma below precisely relates the elements of $A_{i,j}$ to those of $A_{j,i}$.

Lemma 8. *Let $x \in \{0,1\}^n$ be an arbitrary search point, $i > j > 0$ and $A_{i,j}, A_{j,i}$ as in Lemma 7.*

(i) If $Z(x) \leq n/2$ then for every $A \subseteq A_{i,j}$, there exists $B \subseteq A_{j,i}$ such that the following holds. For all $b \in B$, there is $a \in A$ such that $a \geq b$ and $|B| \geq |A|$.

(ii) If $Z(x) \geq n/2$ then for every $B \subseteq A_{j,i}$, there exists $A \subseteq A_{i,j}$ such that the following holds. For all $a \in A$, there is $b \in B$ such that $a \geq b$, and $|A| \geq |B|$.

To derive Theorem 4 from the lemmas above, observe that for $i = 0$ or $j = 0$ the offspring is comparable and therefore (1) evaluates to 0. Then recall that SDBV is chosen so that every offspring which is not comparable to the parent is accepted when $Z(x^t) \leq n/2$, and rejected when $Z(x^t) > n/2$. Hence, for $i, j > 0$ the summands of both sums in (1) are either all $\mathbf{1}_{f(y)>f(x)} - 1$ or all $\mathbf{1}_{f(y)>f(x)}$, and combining the monotonicity of f with Lemma 8 suffices to conclude.

4 SDBV Does Not Maximize Optimization Time

In the previous section we showed that the drift, i.e. the expected amount by which we move *towards* the optimum, is minimized in all points by SDBV (in the class of dynamic monotone functions). It is thus tempting to believe that SDBV is the function for which the expected optimization time is extremal. However, this is not true in general. In this section, we show by direct computation for odd $n \in [9, 45]$ that SDBV is *not* the function that maximizes the runtime. Runtime analysis is usually conducted using drift analysis, but as we have established previously that SDBV minimizes the drift, this approach will not allow us to prove this result. Instead, we compute the expected hitting time of SDBV and other related functions exactly for small values of n. The expected optimization time of a function f when started at some $x \in \{0,1\}^n$ may be expressed as an affine combination of the expected runtimes when started at all other $y \in \{0,1\}^n$. We employ computational methods to solve this system of equations, attaining (exact) numerical values for the expected running time when the problem dimension is small.

Table 1. Drift and expected runtime for the $(1+1)$-EA run on variants of SDBV with different cut-offs for $n = 9$.

(a) Drift in the number of zeros. The drift at all points is minimized by the standard SDBV, as expected from the theoretical analysis done in the previous section. We highlight the places where the drift is higher for different cutoff.

	cutoff 4.5	cutoff 3.5	cutoff 2.5
Δ_0	0	0	0
Δ_1	0.01235	0.01235	0.01235
Δ_2	0.05898	0.05898	0.05898
Δ_3	0.13489	0.13489	**0.16442**
Δ_4	0.23572	**0.24664**	**0.24664**
Δ_5	0.34683	0.34683	0.34683
Δ_6	0.46822	0.46822	0.46822
Δ_7	0.61454	0.61454	0.61454
Δ_8	0.79012	0.79012	0.79012
Δ_9	1	1	1

(b) Expected hitting time. Although the drift is minimized by the 'standard' SDBV, the expected time is not. We highlight the places in which the other variants have higher expected optimization time.

	cutoff 4.5	cutoff 3.5	cutoff 2.5
$E[T_0]$	0	0	0
$E[T_1]$	30.1845	**30.1861**	30.0440
$E[T_2]$	41.2612	**41.2646**	40.9707
$E[T_3]$	47.1214	**47.1276**	46.3839
$E[T_4]$	50.7524	**50.7716**	50.1061
$E[T_5]$	53.3796	**53.3959**	52.7251
$E[T_6]$	55.3045	**55.3210**	54.6501
$E[T_7]$	56.7601	**56.7766**	56.1057
$E[T_8]$	57.8835	**57.9000**	57.2291
$E[T_9]$	58.7644	**58.7809**	58.1100
$E[T]$	50.9855	**50.9997**	50.3553

Intuition: Jumps Are Beneficial. In Sect. 3, we showed that whenever $Z^t < n/2$, a drift minimizing function leads to the acceptance of as many offspring as possible - while maintaining monotonicity at each step, whereas at times t such that $Z^t > n/2$ this is achieved by the rejection of as many offspring as possible. At the point $Z^t = n/2$, both strategies are pessimal in the sense that they lead to the same minimized drift. Recall that we have opted for FDBV in this case. Surprisingly, while the number of potential accepted offspring when $Z^t = n/2$ does not affect the drift, it does play a role for the expected optimization time. It turns out to be beneficial for the optimization of a function to accept more offspring. For an intuitive understanding, consider the following toy example. Assume that for some function f_1 we could only gain progress of one at $Z^t = n/2$ (i.e. $Z^t - Z^{t+1} = 1$), with some probability p, and would stay put otherwise. For a function f_2, we also make progress of one with probability p, but we have the additional options $Z^t - Z^{t+1} = k$ and $Z^t - Z^{t+1} = -k$, both with some probability $q > 0$. Then the drift is identical, but we want to argue that it is still easier to optimize f_2 than f_1. The reason is that progress gets harder as we approach the optimum - evidenced by a drift bound of $\Theta(s^2/n^2)$ proven below. Therefore it is harder to progress from $Z^t = n/2$ to $Z^t = n/2 - k$ than to progress from $Z^t = n/2 + k$ to $Z^t = n/2$. Hence, if we jump k towards the optimum, we gain more (namely, the expected time to proceed from $Z^t = n/2$ to $Z^t = n/2 - k$) than we lose if we jump k away from the optimum (the expected time to proceed from $Z^t = n/2 + k$ to $Z^t = n/2$). Therefore, optimizing f_2 takes less expected time than optimizing f_1. Returning to SDBV, this suggests that we can increase the expected optimization time by expanding the interval in which we use FDBV, thus accepting many offspring in a slightly larger range.

A Simplified Version of SDBV. In the characterization of SBDV employed so far, in principle every search point $x \in \{0,1\}^n$ induces a different function. This would require up to 2^n states to be modeled as a Markov chain and would lead to computational problems even for small problem sizes. One can show that it is enough to consider the number of zeros in the parent, which allows us to analyze an equivalent randomized process that requires only $n+1$ states. The optimization can then be modeled as a Markov chain with the state at time t given by the number of zeros in x^t and the expected optimization time the expected number of steps required to reach state 0.

We proceed to show that the hardest dynamic monotone function to optimize on $(1+1)$-EA is not necessarily the one with lowest drift. For this purpose, we have to compute the hitting times precisely. Thus, all calculations have to be done using fractions without any rounding.[6] We will now look at $n = 9$, as it is the smallest n where the desired effect is observable. By the proof in Sect. 3, SDBV is drift minimizing, which selects by ADBV when $Z(x^t) < 4.5$ and selects by FDBV when $Z(x^t) \geq 4.5$. We compare this with functions with smaller cutoff for switching from ADBV to FDBV. Let $\triangle_s$ denote the drift when the optimization is in state s, that is, when $Z(x^t) = s$. Then reducing the cutoff must increase the drift at the modified states by Theorem 4. We verify this theoretical result for $n = 9$ by computing the drifts $\triangle_s$ numerically in Table 1a. Intuitively, this increase in drift could be expected to be beneficial for faster optimization. However, for a cutoff of 3.5 the expected hitting time increases, see Table 1b. This effect vanishes for an even lower cutoff of 2.5.

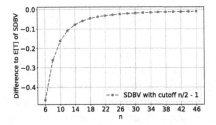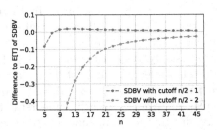

Fig. 1. Change in expected optimization time by reducing the cutoff for even (left) and odd (right) n

We now investigate empirically how the optimal cutoff changes with increasing n. We observe that the parity of n plays a significant role. If n is even, there is a middle state $n/2$, where both ADBV and FDBV selection are drift minimizing. If n is odd for both states $\lceil x/2 \rceil$ and $\lfloor x/2 \rfloor$ the respective drift minimizing strategy is unique. First, consider even n. Figure 1 shows that for small n, even the slightest decrease in the cutoff does not increase the hardness. On the other hand, for all tested odd $n \geq 9$, decreasing the cutoff by 1 to $n/2 - 1$ increases

---

[6] For readability, the final results are then rounded to 4 digits. The code is available at https://github.com/OliverSieberling/SDBV-EA.

the expected hitting time. This effect disappears for cutoff $n/2 - 2$. Modifying the cutoff by some fixed amount changes the behavior only for a fixed amount of states. With increasing n, these modified states become less relevant for the entire optimization. Therefore, it is not surprising that the expected hitting time of SDBV with decreased cutoff approaches that of SDBV as $n \to \infty$.

5 The Expected Optimization Time of SDBV Is $\Theta(n^{3/2})$

In the previous two sections we have established first that SDBV minimizes the drift towards the optimum, and then showed that this does not imply that it maximizes the runtime. However, in this section we will prove that the runtime on SDBV is $\Theta(n^{3/2})$, thus establishing Theorem 6. The upper bound for the runtime follows from the general upper bound of $O(n^{3/2})$ in the PO-EA framework.[7] Since this upper bound holds for *all* dynamic montone functions, it also proves that SDBV gives *asymptotically* the highest runtime. To obtain these results, it thus suffices to prove the following lemma.

Lemma 9. *When started on a uniformly random point $x^0 \in \{0,1\}^n$, the $(1+1)$-EA requires $\Omega(n^{3/2})$ steps in expectation to optimize SDBV.*

Our proof follows that of the lower bound on the runtime of the PO-EA by Colin, Doerr and Férey [2]; as in their paper, we only keep a sketch of the proof due to space limitation. We start by estimating the drift of SDBV at every point. While an upper bound on this drift would be sufficient to derive Lemma 9, we also provide a lower bound estimate. As we have established that SDBV minimizes the drift among dynamic monotone functions, this lower bound holds not only for SDBV, but any such function, which we believe to be interesting on its own.

Lemma 10. *Consider the $(1+1)$-EA on SDBV. Let x^t be the parent at time t, and $\Delta^t = \mathbf{E}\left[Z(x^t) - Z(x^{t+1}) \mid x^t\right]$ the drift at time t. Then for every $l \geq 0$, we have $\Delta^t = \Theta\left((Z(x^t)/n)^2\right)$.*

In addition to the drift at all points, two additional results are needed:

(i) with high probability, at least once during the optimization the number of zero-bits falls in the interval $[\sqrt{n}, 2\sqrt{n}]$;

(ii) while the number of zero-bits is in the interval $[\sqrt{n}, 3\sqrt{n}]$, the *variance* of the number of zero-bit change at each step is $O(1/\sqrt{n})$.

We then argue as follows: consider the first time t_0 when the number of zero-bits falls in the interval $[\sqrt{n}, 2\sqrt{n}]$. At all points until the optimum is reached, or until the number of zeros goes above $3\sqrt{n}$, we have drift $O(1/n)$ and variance $O(1/\sqrt{n})$. Hence, after $\tilde{t} = O(n^{3/2})$ steps, we have moved by an expected amount of[8] $O(\sqrt{n})$ and have a total variance of $\nu = O(n)$. By Kolmogorov's inequality,

---

[7] Actually, the PO-EA$^-$; see [2] for the details.

[8] One should consider $\tilde{t}$ to be chosen as a very small multiple of $n^{3/2}$, so that progress towards the optimum is small in expectation.

the probability of having exited the interval $(0, 3\sqrt{n})$ in fewer than $\tilde{t}$ steps is $\nu/\Omega(\sqrt{n})^2 \leq 1/2$ for an appropriate choice of $\tilde{t}$. Having probability at least $1/2$ of needing $\tilde{t} \sim n^{3/2}$ to exit the interval $(0, 3\sqrt{n})$ immediately implies that the total expected runtime must be at least $\Omega(n^{3/2})$. The full argument uses tools from martingale theory and may be found in the extended version.

6   Simulations

This section aims to provide empirical support to our theoretical results.[9] In particular, we corroborate our theoretical results for the $(1 + 1)$-EA and test them on the static $(1 + \lambda)$-EA and $(1, \lambda)$-EA with $\lambda = 2 \ln n$, and the SA-$(1, \lambda)$-EA. The selection of algorithms is motivated as follows: The $(1 + \lambda)$-EA is the natural extension of the $(1+1)$-EA, where in each generation λ offspring are produced. This allows parallelization. Its non-elitist cousin $(1, \lambda)$ has been shown to escape local optima more efficiently [15,20]. Finally the SA-$(1, \lambda)$-EA allows to dynamically adjust the offspring population size at runtime to the difficulty of the current optimization phase. In this setting, Kaufmann, Larcher, Lengler and Zou formulated the conjecture that ADBV was the hardest dynamic monotone function to optimize. For each algorithm, we investigate problem sizes from $n = 20$ to $n = 420$ in increments of 20. For the SA-$(1, \lambda)$-EA, the update strength is set to $F = 1.15$ and the success ratio to $s = 0.25$. Across all algorithms, the mutation rate is set to $1/n$. We always start with a uniformly random bitstring. The algorithm terminates when the optimum $\{1\}^n$ is found. All plots are averaged over 500 runs. As our benchmarks, we evaluate the following: the building blocks of SDBV, that is ADBV and FDBV, the PO-EA, which for the $(1 + 1)$-EA is asymptotically equivalent to SDBV, and the PO-EA$^-$. We further include DBV and noisy linear functions (with discrete uniform distribution over $[n]$) which are two well-researched dynamic monotone functions, as well as the PO-EA. We refer the interested reader for definitions of these functions and background to [23,31] and [2] respectively. Finally, we include OneMax, the "drosophila" of functions in the study of evolutionary algorithms. As we see in Fig. 2, SDBV and ADBV track each other closely. It is also visible in the algorithms with multiple offspring - both elitist and non-elitist, that asymptotically there is little difference between the SDBV and the ADBV. Differences in the function far away from the optimum play little role asymptotically, even when the selection behavior is the polar opposite, already for moderate problem sizes. This behavior is evident across algorithms, giving credence to the conjecture that SDBV realizes the pessimism of the PO-EA for a large class of algorithms. Up to a slightly smaller multiplicative constant, SDBV indeed matches the runtime even of the fully pessimistic variant of the PO-EA. Finally, we see that the $\Theta(n^{3/2})$ runtime holds for the much larger class of algorithms beyond the $(1+1)$-EA, making it the first known dynamic monotone function with runtime $\omega(n \log n)$ in a setting where they are known to be efficient. In particular, we see a separation between the runtime of dynamic monotone functions such as for Dynamic Binary Value and

---

[9] The code is available at https://github.com/OliverSieberling/SDBV-EA.

Noisy Linear where randomization occurs naively and whose runtime is known to lie in $O(n \log n)$ generations for the regime where optimization is provably efficient [23,31]. I.e., dynamic monotone functions with "uniform" randomization appear intrinsically easier than SDBV.

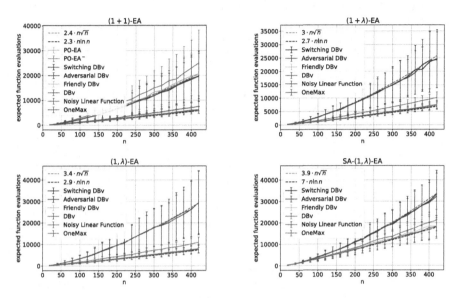

Fig. 2. Runtime comparisons for the $(1 + 1)$-EA, the $(1 + \lambda)$-EA, the $(1, \lambda)$-EA and the SA-$(1, \lambda)$-EA.

7   Conclusion

We have shown that at least in the context of dynamic monotone functions, the PO-EA framework is not overly pessimistic but that there exist indeed functions which pose the difficulties ascribed abstractly to an evolutionary algorithm. We have proved this for the case of the classical $(1 + 1)$-EA and our experiments show that this holds for general offspring sizes, hence also for the $(1 + \lambda)$-EA as well as for some comma-strategies, both the static $(1, \lambda)$-EA and the self-adjusting $(1, \lambda)$-EA require the same asymptotic number of function evaluations of $\Theta(n^{3/2})$. Our function SDBV which materializes this pessimism also minimizes drift. It is however, when conducting a precise runtime analysis for small dimensions, not the function with highest expected optimization time. Which function indeed maximizes expected optimization time is still open. As we see from our experiments, the difference in runtime to the drift minimizing function SDBV decreases for increased dimension. This motivates our concluding conjecture.

Conjecture 11. The expected runtime of the hardest dynamic monotone function for the $(1+1)$-EA exceeds the expected runtime of SDBV at most by $o(n)$ steps.

References

1. Buskulic, N., Doerr, C.: Maximizing drift is not optimal for solving OneMax. Evol. Comput. **29**(4), 521–541 (2021)
2. Colin, S., Doerr, B., Férey, G.: Monotonic functions in EC: anything but monotone! In: Genetic and Evolutionary Computation Conference (GECCO), pp. 753–760 (2014)
3. Corus, D., He, J., Jansen, T., Oliveto, P.S., Sudholt, D., Zarges, C.: On easiest functions for mutation operators in bio-inspired optimisation. Algorithmica **78**(2), 714–740 (2017)
4. Doerr, B., Doerr, C., Yang, J.: Optimal parameter choices via precise black-box analysis. Theor. Comput. Sci. **801**, 1–34 (2020)
5. Doerr, B., Gießen, C., Witt, C., Yang, J.: The $(1+\lambda)$ evolutionary algorithm with self-adjusting mutation rate. Algorithmica **81**, 593–631 (2019)
6. Doerr, B., Jansen, T., Sudholt, D., Winzen, C., Zarges, C.: Optimizing monotone functions can be difficult. In: Schaefer, R., Cotta, C., Kołodziej, J., Rudolph, G. (eds.) PPSN 2010. LNCS, vol. 6238, pp. 42–51. Springer, Heidelberg (2010). https://doi.org/10.1007/978-3-642-15844-5_5
7. Doerr, B., Jansen, T., Sudholt, D., Winzen, C., Zarges, C.: Mutation rate matters even when optimizing monotonic functions. Evol. Comput. **21**(1), 1–27 (2013)
8. Doerr, B., Johannsen, D., Winzen, C.: Multiplicative drift analysis. Algorithmica **64**, 673–697 (2012)
9. Doerr, B., Künnemann, M.: How the $(1+\lambda)$ evolutionary algorithm optimizes linear functions. Theor. Comput. Sci. **561**, 3–23 (2015)
10. Doerr, B., Le, H.P., Makhmara, R., Nguyen, T.D.: Fast genetic algorithms. In: Genetic and Evolutionary Computation Conference (GECCO) (2017)
11. Doerr, B., Lissovoi, A., Oliveto, P.S., Warwicker, J.A.: On the runtime analysis of selection hyper-heuristics with adaptive learning periods. In: Genetic and Evolutionary Computation Conference (GECCO), pp. 1015–1022 (2018)
12. Doerr, C., Janett, D.A., Lengler, J.: Tight runtime bounds for static unary unbiased evolutionary algorithms on linear functions. In: Genetic and Evolutionary Computation Conference (GECCO) (2023)
13. Doerr, C., Wagner, M.: Simple on-the-fly parameter selection mechanisms for two classical discrete black-box optimization benchmark problems. In: Genetic and Evolutionary Computation Conference (GECCO), pp. 943–950 (2018)
14. He, J., Chen, T., Yao, X.: On the easiest and hardest fitness functions. IEEE Trans. Evol. Comput. **19**(2), 295–305 (2014)
15. Hevia Fajardo, M.A., Sudholt, D.: Self-adjusting population sizes for non-elitist evolutionary algorithms: why success rates matter. Algorithmica **86**, 526–565 (2024)
16. Jägersküpper, J., Storch, T.: When the plus strategy outperforms the comma strategy and when not. In: Foundations of Computational Intelligence (FOCI), pp. 25–32. IEEE (2007)
17. Janett, D., Lengler, J.: Two-dimensional drift analysis: optimizing two functions simultaneously can be hard. Theor. Comput. Sci. **971**, 114072 (2023)
18. Jansen, T.: On the brittleness of evolutionary algorithms. In: Stephens, C.R., Toussaint, M., Whitley, D., Stadler, P.F. (eds.) FOGA 2007. LNCS, vol. 4436, pp. 54–69. Springer, Heidelberg (2007). https://doi.org/10.1007/978-3-540-73482-6_4
19. Jansen, T., Jong, K.A.D., Wegener, I.: On the choice of the offspring population size in evolutionary algorithms. Evol. Comput. **13**(4), 413–440 (2005)

20. Jorritsma, J., Lengler, J., Sudholt, D.: Comma selection outperforms plus selection on OneMax with randomly planted optima. In: Genetic and Evolutionary Computation Conference (GECCO) (2023)

21. Kaufmann, M., Larcher, M., Lengler, J., Zou, X.: Self-adjusting population sizes for the $(1, \lambda)$-EA on monotone functions. In: Rudolph, G., Kononova, A.V., Aguirre, H., Kerschke, P., Ochoa, G., Tušar, T. (eds.) PPSN 2022. LNCS, vol. 13399, pp. 569–585. Springer, Cham (2022). https://doi.org/10.1007/978-3-031-14721-0_40

22. Kaufmann, M., Larcher, M., Lengler, J., Zou, X.: OneMax is not the easiest function for fitness improvements. In: Pérez Cáceres, L., Stützle, T. (eds.) EvoCOP 2023. LNCS, vol. 13987, pp. 162–178. Springer, Cham (2023). https://doi.org/10.1007/978-3-031-30035-6_11

23. Kaufmann, M., Larcher, M., Lengler, J., Zou, X.: Self-adjusting population sizes for the $(1, \lambda)$-EA on monotone functions. Theor. Comput. Sci. **979**, 114181 (2023)

24. Lässig, J., Sudholt, D.: General scheme for analyzing running times of parallel evolutionary algorithms. In: Schaefer, R., Cotta, C., Kołodziej, J., Rudolph, G. (eds.) PPSN 2010. LNCS, vol. 6238, pp. 234–243. Springer, Cham (2010). https://doi.org/10.1007/978-3-642-15844-5_24

25. Lässig, J., Sudholt, D.: Adaptive population models for offspring populations and parallel evolutionary algorithms. In: Foundations of Genetic Algorithms (FOGA), pp. 181–192 (2011)

26. Lengler, J.: A general dichotomy of evolutionary algorithms on monotone functions. IEEE Trans. Evol. Comput. **24**(6), 995–1009 (2019)

27. Lengler, J.: Drift analysis. In: Doerr, B., Neumann, F. (eds.) Theory of Evolutionary Computation. NCS, pp. 89–131. Springer, Cham (2020). https://doi.org/10.1007/978-3-030-29414-4_2

28. Lengler, J., Martinsson, A., Steger, A.: When does hillclimbing fail on monotone functions: an entropy compression argument. In: Analytic Algorithmics and Combinatorics (ANALCO), pp. 94–102. SIAM (2019)

29. Lengler, J., Meier, J.: Large population sizes and crossover help in dynamic environments. In: Bäck, T., et al. (eds.) PPSN 2020. LNCS, vol. 12269, pp. 610–622. Springer, Cham (2020). https://doi.org/10.1007/978-3-030-58112-1_42

30. Lengler, J., Riedi, S.: Runtime analysis of the $(\mu + 1)$-EA on the dynamic BinVal function. In: Zarges, C., Verel, S. (eds.) EvoCOP 2021. LNCS, vol. 12692, pp. 84–99. Springer, Cham (2021). https://doi.org/10.1007/978-3-030-72904-2_6

31. Lengler, J., Schaller, U.: The $(1 + 1)$-EA on noisy linear functions with random positive weights. In: Symposium Series on Computational Intelligence (SSCI), pp. 712–719. IEEE (2018)

32. Lengler, J., Steger, A.: Drift analysis and evolutionary algorithms revisited. Comb. Probab. Comput. **27**(4), 643–666 (2018)

33. Lengler, J., Zou, X.: Exponential slowdown for larger populations: the $(\mu + 1)$-EA on monotone functions. Theor. Comput. Sci. **875**, 28–51 (2021)

34. Neumann, F., Oliveto, P.S., Witt, C.: Theoretical analysis of fitness-proportional selection: landscapes and efficiency. In: Proceedings of the 11th Annual Conference on Genetic and Evolutionary Computation, pp. 835–842 (2009)

35. Rowe, J.E., Sudholt, D.: The choice of the offspring population size in the $(1, \lambda)$ evolutionary algorithm. Theor. Comput. Sci. **545**, 20–38 (2014)

36. Sudholt, D.: A new method for lower bounds on the running time of evolutionary algorithms. IEEE Trans. Evol. Comput. **17**(3), 418–435 (2012)

37. Witt, C.: Tight bounds on the optimization time of a randomized search heuristic on linear functions. Comb. Probab. Comput. **22**(2), 294–318 (2013)

A Theoretical Investigation of Termination Criteria for Evolutionary Algorithms

Jonathan E. Rowe[✉]

School of Computer Science, University of Birmingham, Birmingham B15 2TT, UK
J.E.Rowe@cs.bham.ac.uk

Abstract. We take a theoretical approach to analysing conditions for terminating evolutionary algorithms. After looking at situations where much is known about the particular algorithm and problem class, we consider a more generic approach. Schemes that depend purely on the previous time to improvement are shown not to work. An alternative criterion, the λ-parallel scheme, does terminate correctly (with high probability) for any randomised search heuristic algorithm on any problem, provided certain conditions on the improvement probabilities are met. A more natural and less costly approach is then presented based on the runtime so far. This is shown to work for the classes of monotonic and path problems (for Randomised Local Search). It remains an open question whether it works in a more general setting.

1 Introduction

When analysing the runtime of evolutionary algorithms mathematically, it is common to make an assumption that the main loop which governs the algorithm runs forever. Theoretical results are then proved about the expected number of iterations to achieve the optimum (or some other target state). Of course, in practice one writes algorithms with a termination criterion on the loop. There have been a number of empirical studies of various heuristics that could be used (see for example [8–10]), but it is interesting to ask if there is any theoretical guidance as to how to do this. We hope to find generic stopping criteria that satisfy the following:

1. Criteria do not depend on the details of the problem being optimised (since we want our evolutionary algorithms to be generic problem solvers).
2. They depend as little as possible on the details of the algorithm being used (as we would like to reuse the ideas in different evolutionary algorithms).
3. They should allow the problem to be solved with high probability, while not increasing the runtime unduly (e.g. by more than a constant factor).

We will begin by looking at what can be done when we do know some details about the expected performance of our algorithm on a problem class. We will then study the effectiveness of some potential generic criteria. First we will look at schemes based on the previous improvement time, and show that these do not

© The Author(s), under exclusive license to Springer Nature Switzerland AG 2024
T. Stützle and M. Wagner (Eds.): EvoCOP 2024, LNCS 14632, pp. 162–176, 2024.
https://doi.org/10.1007/978-3-031-57712-3_11

succeed. This leads us to a more artificial criterion, where we re-run improvement steps several times to get better estimates of the improvement time. This can be made to work, but at a cost. We then consider looking back over several improvement stages, and propose a scheme where we terminate if we have seen no improvements after a time equal to the entire runtime to achieve the previous improvement. This does work on some problems, but it is an open question how general this is. We will illustrate the ideas using the following examples (which will also provide counter-examples for some cases). In each case, we will consider the application of termination criteria to Random Local Search (RLS), which is defined as Algorithm 1. Note that for the purposes of these examples, we always start at 0^n rather than at a random initial string.

Path problems. A *path* problem is defined by a path on the hypercube $\{0, 1\}^n$ starting at 0^n and finishing at a target string z, such that any two points on the path always have distance at least 2 from each other, except for consecutive points. The fitness of bitstrings on the path is given by the distance travelled along the path. Since we start at 0^n and only accept improving moves, we do not need to consider the fitness of points off the path, which we assume is much worse. For RLS on path problems, the probability of improvement at each stage is $1/n$. In this case, the probability of improvement remains constant until the target is found.

Monotonic problems. A problem is *monotonic* if changing any 0 to a 1 always improves the fitness. Linear functions (such as ONEMAX) are monotonic. For RLS on monotonic functions, if there are k zeroes in a string, then the probability of improvement is k/n. In this case, the probability of improvement gets smaller as the search progresses.

2   Useful Inequalities

We make use of the following well known inequalities:

$$(1 + x)^y \geq 1 + xy, \text{ for all } x \geq -1, y \geq 1,$$
$$(1 + x) \leq \exp(x), \text{ for all } x,$$
$$(1 - x) \geq \exp(-2x), \text{ for all } 0 \leq x \leq 1/2,$$
$$\left(\frac{n}{k}\right)^k \leq \binom{n}{k} \leq \left(\frac{en}{k}\right)^k, \text{ for all } 1 \leq k \leq n.$$

We also use the following inequality from [11]:

$$(1 - x)^y \leq \frac{1}{1 + xy}, \text{ for all } 0 \leq x \leq 1, y \geq 0.$$

These, and many other useful results, are collected together in [5].

```
Let x = 0ⁿ ;
while Termination criterion not satisfied do
    Choose i ∈ {1,...,n} uniformly at random ;
    Let y = x ⊕ eᵢ ;
    if f(y) ≥ f(x) then
        | Let x = y ;
    end
end
```

Algorithm 1: Random Local Search (RLS), starting at the all zeros string. Note that e_i indicates the string containing a 1 at position i and zeros elsewhere.

3 Known Expected Runtime

When we do have access to details of the problem class and the specific algorithm being used, we can use this knowledge to define problem-specific stopping criteria. If we have an upper bound on the expected running time which has tight tail bounds, then we can use those. For example, if an algorithm follows a multiplicative drift (with factor δ), and has a maximum distance n to the optimum, then running for $((c+1)\log n)/\delta$ again ensures we find the optimum with probability at least $1 - 1/n^c$ (see [6]).

In cases where we have an estimate for the expected runtime, but no tail bounds, we can use the method of "independent phases" [1]. If τ is an upper bound on the expected runtime, $E[T]$, independent of the starting point, then Markov's inequality tells us that

$$\Pr(T > 2\tau) \leq \frac{E[T]}{2\tau} \leq \frac{1}{2}.$$

We run the algorithm for k phases, each of length 2τ. Since the upper bound on the runtime is independent of the starting point we can consider each phase as an independent Bernoulli trial. The probability of failing to find the optimum in all k phases is thus at most $1/2^k$. So we have:

$$\Pr(T > 2k\tau) \leq \frac{1}{2^k}.$$

However, it can happen that we have a theoretical expression for the expected runtime, but this involves parameters which the algorithm designer does not know. One example is the analysis of simulated annealing on the minimum spanning tree problem, where the runtime depends on the maximum and minimum edge weights [4].

A more challenging situation is the hidden subset problem, where the fitness function depends on only a (possibly small) unknown subset of the bits [2]. Here we have algorithms which solve the problem in a time that depends only on the number of hidden bits, rather than the full string length [3], but given this number is unknown it is hard to see how to establish a termination criterion for such algorithms. An alternative approach to this problem makes use of an adaptive mutation strategy (see Algorithm 4 of [7]) which does allow the value n to be estimated as the search progresses.

4 Known Improvement Probability

If we have a lower bound on the probability that the algorithm makes progress at each step, then we can use this to determine a stopping criterion. Let p be such a lower bound. Then we will stop if we have had $(2/p)\log(\ell)$ steps without any improvement (where ℓ is a parameter to be determined). The probability that this fails on a single step (that is, we stop when we should not have) is at most

$$(1-p)^{(2/p)\log\ell} \leq \frac{1}{\ell^2}.$$

By the union bound, the probability that the test fails at least once over ℓ improving steps is at most $1/\ell$. We can see that the method will work with high probability if paths from the starting point to the optimum are at most ℓ in length.

For example, consider RLS running on a path problem. The probability that RLS makes an improvement at any step is $1/n$. So by setting the termination condition to $2n\log(\ell)$ we can successfully optimise path problems with path length up to ℓ. Since the expected time to find the optimum is $O(n\ell)$ we see that the final phase when we are waiting to (correctly) terminate is not asymptotically significant.

It is unfortunate that to make this work, we need a good idea what the path length is. We can obviate this by setting a termination criterion that stops if n^2 steps occur without improvement. Now the probability of failing on a single step is

$$(1-1/n)^{n^2} \leq \exp(-n)$$

which means that we will succeed with high probability for any path problem (since the path length must be less than 2^n). This is a trivial price to pay if the path length is quadratic or longer, but it is expensive for sub-quadratic path length problem instances.

5 Towards a Generic Adaptive Termination Criterion

5.1 Looking at the Previous Improvement Time

Ideally, we would like a termination criterion where we do not have to estimate in advance the probability of an improvement. One idea is to observe that for a lot of problem classes, evolutionary algorithms make fairly steady progress throughout the run. The expected time it takes to get an improvement is not excessively longer that the expected time to get the previous improvement. This suggests an adaptive, online approach where we decide it is time to terminate if we are taking too long to improve compared to the previous step.

1. Run the algorithm until there is an improvement.
2. Let t be the time taken to achieve this improvement.
3. Run the algorithm until a new improvement is found, or $h(t)$ iterations have occurred.
4. If a new improvement has been found in less than $h(t)$ iterations, then go to step 2.
5. Otherwise terminate.

5.2   Path Problems

We can try this idea out on RLS working on path problems since in this case, each improvement is governed by identical independent geometric distributions, with success probability $1/n$. Indeed, this would seem to be a best case for the proposal, since more commonly it takes longer and longer to get improvements, and therefore more chance that we might not wait long enough. So let us say that we will terminate if, following an improvement that took t iterations, we have now waited $h(t) = ct$ iterations with no further improvement (for some constant c). Unfortunately, as the next proposition shows, the probability that this fails is asymptotically a constant, $1/(c+1)$. Therefore the chance of at least one fail in finding all the required improvements (of which ℓ, the path length, are required) is exponentially close to 1 (as a function of ℓ).

Proposition 1. *Let X, Y be independent random geometric variables, with success probability $1/n$, and let $c \geq 1$. Then*

$$\frac{1}{c+1} - \left(\frac{c}{c+1}\right)\frac{1}{n-1} \leq \Pr(Y > cX) \leq \frac{1}{c+1} + \left(\frac{1}{c+1}\right)\frac{2n-1}{(n-1)^2}.$$

Proof. Note that:

$$\Pr(Y > cX) = \sum_{t=1}^{\infty} \frac{1}{n}\left(1 - \frac{1}{n}\right)^{t-1}\left(1 - \frac{1}{n}\right)^{\lfloor ct \rfloor}$$

For a lower bound, we have

$$Pr(Y > cX) = \sum_{t=1}^{\infty} \frac{1}{n}\left(1 - \frac{1}{n}\right)^{t-1}\left(1 - \frac{1}{n}\right)^{\lfloor ct \rfloor} \geq \sum_{t=1}^{\infty} \frac{1}{n}\left(1 - \frac{1}{n}\right)^{t-1}\left(1 - \frac{1}{n}\right)^{ct}$$

$$= \frac{1}{n-1}\sum_{t=1}^{\infty}\left(1 - \frac{1}{n}\right)^{(c+1)t} = \frac{1}{n-1}\left(\frac{1}{1 - (1 - 1/n)^{c+1}} - 1\right).$$

Using

$$\left(1 - \frac{1}{n}\right)^{c+1} \geq 1 - \frac{c+1}{n}$$

we have

$$Pr(Y > cX) \geq \frac{1}{n-1}\left(\frac{n}{c+1} - 1\right) = \frac{1}{c+1} - \left(\frac{c}{c+1}\right)\frac{1}{n-1}.$$

For the upper bound we have

$$Pr(Y > cX) = \sum_{t=1}^{\infty} \frac{1}{n}\left(1 - \frac{1}{n}\right)^{t-1}\left(1 - \frac{1}{n}\right)^{\lfloor ct \rfloor} \leq \sum_{t=1}^{\infty} \frac{1}{n}\left(1 - \frac{1}{n}\right)^{t-1}\left(1 - \frac{1}{n}\right)^{ct-1}$$

$$= \frac{n}{(n-1)^2}\left(\frac{1}{1 - (1 - 1/n)^{c+1}} - 1\right).$$

Using

$$\left(1 - \frac{1}{n}\right)^{c+1} \leq \frac{1}{1 + (c+1)/n}$$

we have

$$Pr(Y > cX) \leq \frac{n}{(n-1)(c+1)} = \frac{1}{c+1} + \left(\frac{1}{c+1}\right)\frac{2n-1}{(n-1)^2}.$$

$\square$

Given the discussion of the previous section, one might think we need to wait $ct \log t$ steps or even t^2 steps before terminating. In fact we will show that *any* polynomial in t is insufficient. Consider the case where we wait $h(t) = ct^k$ for some constants c and k. The probability that we stop prematurely at one particular stage is

$$\sum_{t=1}^{\infty} \frac{1}{n}\left(1 - \frac{1}{n}\right)^{t-1}\left(1 - \frac{1}{n}\right)^{ct^k} = \frac{1}{n-1}\sum_{t=1}^{\infty}\left(1 - \frac{1}{n}\right)^{ct^k+t}$$

$$\geq \frac{1}{n-1}\sum_{t=1}^{\infty}\left(1 - \frac{1}{n}\right)^{(c+1)t^k} \geq \frac{1}{n-1}\sum_{t=1}^{\infty} \exp\left(-\frac{2(c+1)t^k}{n}\right)$$

$$\geq \frac{1}{n-1}\sum_{t=1}^{\lfloor n^{1/k} \rfloor} \exp\left(-\frac{2(c+1)t^k}{n}\right) \geq \frac{1}{n-1}\sum_{t=1}^{\lfloor n^{1/k} \rfloor} \exp\left(-\frac{2(c+1)\lfloor n^{1/k} \rfloor^k}{n}\right)$$

$$\geq \frac{1}{n-1}\sum_{t=1}^{\lfloor n^{1/k} \rfloor} \exp(2(c+1)) = \exp(2(c+1))\frac{\lfloor n^{1/k} \rfloor}{n-1}$$

$$\geq \exp(2(c+1))\frac{n^{1/k} - 1}{n-1} \geq \frac{\exp(2(c+1))}{2}\frac{1}{n^{1-1/k}}.$$

Consequently, the probability of a success on a single improvement is at most

$$1 - \left(\frac{\exp(2(c+1))}{2}\frac{1}{n^{1-1/k}}\right).$$

On path problems, we need ℓ improvements to solve the problem. The probability of this number happening without premature termination is therefore at most

$$\left(1 - \left(\frac{\exp(2(c+1))}{2}\frac{1}{n^{1-1/k}}\right)\right)^{\ell} \leq \exp\left(-\frac{\exp(2(c+1))}{2}\frac{\ell}{n^{1-1/k}}\right).$$

So if $\ell = \Omega(n)$, then the probability of success goes to zero as $n \to \infty$.

5.3   Monotonic Problems

In the case of RLS on monotonic problems the expected time for an improvement increases at each stage of the search process. This means that the previous waiting time is less likely to be a good estimate for the current waiting time. We again show that taking $h(t)$ to be any polynomial functions is insufficient. If there are z zeros remaining, then the success probability is z/n. Taking $h(t) = ct^k$, we have the probability that the termination scheme fails at this stage is:

$$\sum_{t=1}^{\infty}\left(\frac{z+1}{n}\right)\left(1-\frac{z+1}{n}\right)^{t-1}\left(1-\frac{z}{n}\right)^{ct^k} \geq \left(\frac{z+1}{n}\right)\left(1-\frac{z}{n}\right)^{c}.$$

Therefore, the probability that the scheme succeeds overall is at most

$$\prod_{z=1}^{n-1}\left(1-\left(\frac{z+1}{n}\right)\left(1-\frac{z}{n}\right)^{c}\right) \leq \prod_{z=n/4-1}^{3n/4}\left(1-\left(\frac{z+1}{n}\right)\left(1-\frac{z}{n}\right)^{c}\right)$$

$$\leq \prod_{z=n/4-1}^{3n/4}\left(1-\frac{1}{4}\left(1-\frac{3}{4}\right)^{c}\right) = \prod_{z=n/4-1}^{3n/4}\left(1-\frac{1}{2^{2c+2}}\right)$$

$$\leq \left(1-\frac{1}{2^{2c+2}}\right)^{n/2} \leq \exp\left(-\frac{n}{2^{2c+3}}\right)$$

which goes to zero as $n \to \infty$.

6   Better Estimations of the Improvement Time

6.1 λ-Parallel Schemes

One of the reasons just using the previous time to improvement as a criterion is problematic, is that there is a good chance of occasionally seeing very quick successes, which would then lead to an unrealistically short waiting time for the next improvement. So we need to try to get a better estimate of the mean improvement time. This suggests the idea of re-running the algorithm several times from the same starting point, and measuring the time to the next improvement in each case. This gives us the following λ-parallel termination scheme:

1. Run λ copies of the algorithm from the same starting point until each one has an improvement.

2. Let t be the total of those λ improvement times.
3. Now run λ copies of the algorithm starting from the best point now reached.
4. If all of these copies fail to find an improvement within $h(t)$ steps, then terminate.
5. Otherwise continue until they have all found an improvement and go to 2.

where $h(t)$ is to be specified. The most obvious choice is to use the mean of the λ improvement times, that is $h(t) = ct/\lambda$ for some constant c.

If we let T be the random variable which is the total time taken to get λ improvements in one round, then for RLS on path problems,

$$\Pr(T = t) = \binom{t-1}{\lambda-1} \left(\frac{1}{n}\right)^{\lambda} \left(1 - \frac{1}{n}\right)^{t-\lambda}$$

since the final step is the time when the last copy succeeds, but the other $\lambda - 1$ copies will have succeeded at different times within the preceding $t - 1$ steps. This means that (for RLS on path problems) the probability that all the next λ copies fail to find an improvement in time $h(t) = ct/\lambda$ is:

$$\sum_{t=\lambda}^{\infty} \binom{t-1}{\lambda-1} \left(\frac{1}{n}\right)^{\lambda} \left(1 - \frac{1}{n}\right)^{t-\lambda} \left(\left(1 - \frac{1}{n}\right)^{ct/\lambda}\right)^{\lambda}$$

$$= \left(\frac{1}{n-1}\right)^{\lambda} \sum_{t=\lambda}^{\infty} \binom{t-1}{\lambda-1} \left(1 - \frac{1}{n}\right)^{(c+1)t}$$

$$\geq \left(\frac{1}{n-1}\right)^{\lambda} \left(\frac{1}{\lambda-1}\right)^{\lambda-1} \sum_{t=\lambda}^{\infty} (t-1)^{\lambda-1} \left(1 - \frac{1}{n}\right)^{(c+1)t}$$

$$\geq \left(\frac{1}{n-1}\right)^{\lambda} \left(\frac{1}{\lambda-1}\right)^{\lambda-1} \sum_{t=\lambda}^{\infty} (t-1)^{\lambda-1} \exp\left(-\frac{2(c+1)t}{n}\right)$$

$$\geq \left(\frac{1}{n-1}\right)^{\lambda} \left(\frac{1}{\lambda-1}\right)^{\lambda-1} \sum_{t=n}^{2n} (t-1)^{\lambda-1} \exp\left(-\frac{2(c+1)t}{n}\right)$$

$$\geq \left(\frac{1}{n-1}\right)^{\lambda} \left(\frac{1}{\lambda-1}\right)^{\lambda-1} \sum_{t=n}^{2n} (n-1)^{\lambda-1} \exp(-4(c+1))$$

$$= \left(\frac{1}{n-1}\right) \left(\frac{1}{\lambda-1}\right)^{\lambda-1} (n+1) \exp(-4(c+1))$$

$$\geq \left(\frac{1}{\lambda-1}\right)^{\lambda-1} \exp(-4(c+1)).$$

Since this is a constant (with regards n and the path length ℓ) the probability of success on ℓ trials is exponential small in ℓ.

We could also try estimating the median improvement time from the λ trials. We could do this by looking at the maximum time, M, required to find an improvement. The chance that this is less than the true median is $1/2^{\lambda}$. If we condition on M exceeding the true median, then the probability that all of the next λ trials fail to improve in M time steps must be less than $1/2^\lambda$. We see that we can control the probability of failure by taking sufficiently large λ. However, for any fixed choice of λ, the failure probability will still be constant with respect to the number of improvements required. This will lead to failure over all, with high probability. Indeed, observing that the maximum of λ trials is bounded above by the sum of the λ trials, we have the probability of failure on a single improvement phase being at least

$$\sum_{t=\lambda}^{\infty} \binom{t-1}{\lambda-1} \left(\frac{1}{n}\right)^\lambda \left(1-\frac{1}{n}\right)^{t-\lambda} \left(\left(1-\frac{1}{n}\right)^t\right)^\lambda$$

which is the same as we had for the mean estimate, taking $c = \lambda$. We therefore have the probability of failure in this case being at least:

$$\left(\frac{1}{\lambda-1}\right)^{\lambda-1} \exp(-4(\lambda+1)).$$

This leads us to consider strengthening the termination criterion by considering different functions $h(t)$. Unfortunately, we now show that a λ-parallel scheme cannot work for RLS on any monotonic function regardless of the choice of $h(t)$, if λ is constant. The problem is that it is too easy for all λ trials to succeed at the very first attempt (giving $t = \lambda$). Suppose there are currently z zeros. The probability that the criterion fails at this stage is:

$$\sum_{t=\lambda}^{\infty} \binom{t-1}{\lambda-1} \left(\frac{z+1}{n}\right)^\lambda \left(1-\frac{z+1}{n}\right)^{t-\lambda} \left(1-\frac{z}{n}\right)^{\lambda f(t)} \geq \left(\frac{z+1}{n}\right)^\lambda \left(1-\frac{z}{n}\right)^{\lambda f(\lambda)}.$$

Using a similar argument to before, the probability that the termination criterion succeeds over the whole run is at most

$$\prod_{z=1}^{n-1} \left(1 - \left(\frac{z+1}{n}\right)^\lambda \left(1-\frac{z}{n}\right)^{\lambda h(\lambda)}\right) \leq \prod_{z=n/4-1}^{3n/4} \left(1 - \left(\frac{z+1}{n}\right)^\lambda \left(1-\frac{z}{n}\right)^{\lambda h(\lambda)}\right)$$

$$\leq \prod_{z=n/4-1}^{3n/4} \left(1 - \left(\frac{1}{4}\right)^\lambda \left(1-\frac{3}{4}\right)^{\lambda h(\lambda)}\right) = \prod_{z=n/4-1}^{3n/4} \left(1 - \frac{1}{2^{2\lambda h(\lambda)+2\lambda}}\right)$$

$$\leq \left(1 - \frac{1}{2^{2\lambda h(\lambda)+2\lambda}}\right)^{n/2} \leq \exp\left(-\frac{n}{2^{2\lambda h(\lambda)+2\lambda+1}}\right)$$

which goes to zero as $n \to \infty$.

6.2   Success with High Probability

We have seen that it is insufficient to take λ resamples of the previous improvement time to inform our termination criterion, regardless of the choice of $h(t)$, as long as λ is a constant. We therefore now consider varying λ. In this section, we will vary λ as a function of the desired success probability—that is, the probability that we do not terminate prematurely when there is still an improvement to be made. We will consider the consequences of this choice later. We also wish our scheme to work for all "reasonable" fitness functions, not just path and monotonic problems. Consequently, we now consider a general situation in which the current probability of finding an improvement is p, and the probability of finding the improvement at the previous stage is q. We will assume that at each stage $q \geq p$ so the problem gets harder (or at least no easier) as the search progresses. However, we will need to assume that q can't be too much greater than p, so that we have a chance of getting a reasonable estimate. So we will assume there is some constant $\alpha \geq 1$ such that $q \leq \alpha p$. For example, for RLS on path problems, we can take $\alpha = 1$ and on monotonic problems we have $\alpha = 2$. We emphasise, though, that we are now in a quite general setting with regards to the algorithm and problem class. We will need the following lemma:

Lemma 1. Let $\lambda > 1$ be an integer and $0 < x < 1$. Then

$$\sum_{t=\lambda}^{\infty} \binom{t-1}{\lambda-1} (1-x)^{(\lambda+1)t} \leq \frac{1}{x^\lambda} \left(\frac{e}{\lambda-1} \right)^{\lambda-1}.$$

Proof.

$$\sum_{t=\lambda}^{\infty} \binom{t-1}{\lambda-1} (1-x)^{(\lambda+1)t} \leq \left(\frac{e}{\lambda-1} \right)^{\lambda-1} \sum_{t=\lambda}^{\infty} (t-1)^{\lambda-1} (1-x)^{(\lambda+1)t}$$

$$\leq \left(\frac{e}{\lambda-1} \right)^{\lambda-1} \sum_{t=\lambda}^{\infty} (t-1)^{\lambda-1} \left(\frac{1}{1+tx} \right)^{\lambda+1}$$

$$\leq \frac{1}{x^{\lambda+1}} \left(\frac{e}{\lambda-1} \right)^{\lambda-1} \sum_{t=\lambda}^{\infty} \left(\frac{t-1}{t+1/x} \right)^{\lambda-1} \left(\frac{1}{1/x+t} \right)^2$$

$$\leq \frac{1}{x^{\lambda+1}} \left(\frac{e}{\lambda-1} \right)^{\lambda-1} \sum_{t=\lambda}^{\infty} \left(\frac{1}{1/x+t} \right)^2$$

$$\leq \frac{1}{x^{\lambda+1}} \left(\frac{e}{\lambda-1} \right)^{\lambda-1} \int_0^\infty \frac{dt}{(1/x+t)^2} = \frac{1}{x^\lambda} \left(\frac{e}{\lambda-1} \right)^{\lambda-1}.$$

$\square$

Theorem 1. If $p \leq q$ and $q \leq \alpha p$ for some constant $\alpha \geq 1$, then the probability that the λ-parallel termination criterion, with $h(t) = t$, fails at any particular stage is at most

$$(4\alpha)^\lambda \left(\frac{e}{\lambda-1} \right)^{\lambda-1}.$$

Proof. We first consider the case where $q \geq p \geq 1/(2\alpha)$. The probability of failure is

$$\sum_{t=\lambda}^{\infty} \binom{t-1}{\lambda-1} q^\lambda (1-q)^{t-\lambda} (1-p)^{\lambda t} \leq \sum_{t=\lambda}^{\infty} \binom{t-1}{\lambda-1} (1-p)^{(\lambda+1)t-\lambda}$$

$$\leq \sum_{t=\lambda}^{\infty} \binom{t-1}{\lambda-1} \left(1-\frac{1}{2\alpha}\right)^{(\lambda+1)t-\lambda} \leq \left(\frac{2\alpha}{2\alpha-1}\right)^\lambda \sum_{t=\lambda}^{\infty} \binom{t-1}{\lambda-1} \left(1-\frac{1}{2\alpha}\right)^{(\lambda+1)t}$$

$$\leq \left(\frac{4\alpha^2}{2\alpha-1}\right)^\lambda \left(\frac{e}{\lambda-1}\right)^{\lambda-1} \leq (4\alpha)^\lambda \left(\frac{e}{\lambda-1}\right)^{\lambda-1}$$

(where we have used Lemma 1). We now consider the case where $p \leq 1/(2\alpha)$. The probability of failure is

$$\sum_{t=\lambda}^{\infty} \binom{t-1}{\lambda-1} q^\lambda (1-q)^{t-\lambda}(1-p)^{\lambda t} = \left(\frac{q}{1-q}\right)^\lambda \sum_{t=\lambda}^{\infty} \binom{t-1}{\lambda-1}(1-q)^t(1-p)^{\lambda t}$$

$$\leq \left(\frac{q}{1-q}\right)^\lambda \sum_{t=\lambda}^{\infty} \binom{t-1}{\lambda-1}(1-p)^{(\lambda+1)t} \leq \frac{1}{p^\lambda} \left(\frac{q}{1-q}\right)^\lambda \left(\frac{e}{\lambda-1}\right)^{\lambda-1}$$

(using Lemma 1)

$$\leq \frac{1}{p^\lambda} \left(\frac{\alpha p}{1-\alpha p}\right)^\lambda \left(\frac{e}{\lambda-1}\right)^{\lambda-1} = \left(\frac{\alpha}{1-\alpha p}\right)^\lambda \left(\frac{e}{\lambda-1}\right)^{\lambda-1}$$

$$\leq \left(\frac{\alpha}{1-1/2}\right)^\lambda \left(\frac{e}{\lambda-1}\right)^{\lambda-1} = (2\alpha)^\lambda \left(\frac{e}{\lambda-1}\right)^{\lambda-1} \leq (4\alpha)^\lambda \left(\frac{e}{\lambda-1}\right)^{\lambda-1}.$$

$\square$

Since we are after a termination criterion that could be used in practice on functions $f : \{0,1\}^n \to \mathbb{R}$, we might be satisfied if our scheme works with high probability over a polynomial (in n) number of improvement steps. Choosing $\lambda = \ln n + 1$ we have, for any constant $c \geq 1$,

$$\left(\frac{4\alpha e^{c+1}}{\lambda-1}\right) = \left(\frac{4\alpha e^{c+1}}{\ln n}\right) \leq 1$$

for sufficiently large n. Hence

$$\left(\frac{4\alpha e}{\lambda-1}\right) \leq \exp(-c)$$

and so the probability the λ-parallel scheme fails at any stage is no more than

$$(4\alpha)^\lambda \left(\frac{e}{\lambda-1}\right)^{\lambda-1} = 4\alpha \left(\frac{4\alpha e}{\lambda-1}\right)^{\lambda-1} = 4\alpha \left(\frac{4\alpha e}{\ln n}\right)^{\ln n} \leq 4\alpha \exp(-c \ln n) = \frac{4\alpha}{n^c}.$$

Since this holds for any constant c (for sufficiently large n) we see that the scheme will succeed over any polynomial number of improvement steps. Of course, it has the disadvantage of requiring an increase in runtime of a factor $\ln n$.

7   Using Several Previous Generations

7.1 λ-Sequential Schemes

From an optimisation point of view, it seems a little perverse to insist on re-running each improvement step multiple times. Instead, we could look at the last λ improvements, and use these to inform our termination criterion. The general λ-sequential scheme is as follows:

1. Run the algorithm until λ improvements have been made.
2. Let t be the total of the previous λ improvement times.
3. Run the algorithm until a new improvement is found, or $h(t)$ iterations have occurred.
4. If a new improvement has been found in less than $h(t)$ iterations, go to 2.
5. Otherwise terminate.

This could be problematic in a couple of ways. First, this does not give us any stopping criterion for the first λ improvements, so one has to be very confident that these will happen (and in a time one is willing to wait). Second, in many situations, such as RLS on monotonic problems, the expected time to improvement increases as the search progresses, and so looking back over λ improvements may not give as accurate information about the current waiting time compared to λ copies of the previous improvement time. This latter consideration is not an issue for RLS on path problems, as the chance of improvement is identically distributed throughout the optimisation process. In this case, we can take $h(t) = \lambda t$ with $\lambda = \log n$, and the scheme will succeed with high probability for paths of any polynomial length. However, this could be an issue for RLS on monotonic problems, and it remains an open question whether this can be made to work.

7.2   Using the Entire Runtime

Instead of just looking at the previous λ improvement times, we now propose to use the entire runtime needed to achieve all the improvements found so far. This gives us the *entire runtime* termination scheme:

1. Run the algorithm until λ improvements have been made.
2. Let t be the total time taken so far.
3. Run the algorithm until a new improvement is found, or $h(t)$ iterations have occurred.
4. If a new improvement has been found in less than $h(t)$ iterations, go to 2.
5. Otherwise terminate.

We now show that this scheme can work for RLS on path problems and monotonic problems with $h(t) = ct$, for suitable choice of constant c. This choice has the advantage that the overall expected runtime is only a constant factor times the expected time to find the optimum. Showing that it can work more generally is an open problem.

For RLS on path problems, we make use of the following lemma (using equation (1.10.49) of [5]):

Lemma 2. *Let* $X_1, \ldots, X_k$ *be random geometric variables with identical success probability* p. *Let* $X = \sum_{i=1}^{k} X_k$ *which has mean* $\mu = k/p$. *Then*

$$\Pr(X \leq \mu/2) \leq \exp(-3k/16).$$

Now we have

Theorem 2. *The probability that the entire runtime termination criterion with* $h(t) = t$ *fails for RLS on a path problem (with any length path) is no more than* $10\exp(-3\lambda/16)$, *where* λ *is the initial number of unchecked improvements.*

Proof. If we have currently seen k improvements, which were found after a total of t iterations, then

$$\Pr(\text{fail at stage } k) = \Pr(\text{fail at stage } k \mid t < nk/2) \Pr(t < nk/2)$$
$$+ \Pr(\text{fail at stage } k \mid t \geq nk/2) \Pr(t \geq nk/2)$$

$$\leq \exp\left(-\frac{3k}{16}\right) + \left(1 - \frac{1}{n}\right)^{nk/2} \leq \exp\left(-\frac{3k}{16}\right) + \exp\left(-\frac{k}{2}\right) \leq 2\exp\left(-\frac{3k}{16}\right).$$

Therefore, by the union bound, the probability of at least one failure on the entire path of length ℓ, starting after the λth improvement is no more than

$$2 \sum_{k=\lambda+1}^{\ell} \exp\left(-\frac{3k}{16}\right) \leq 2 \sum_{k=\lambda+1}^{\infty} \exp\left(-\frac{3k}{16}\right) \leq 10\exp(-3\lambda/16).$$

$\square$

We see that choosing $\lambda = (16/3)\ln n$ gives us an overall failure rate of at most $10/n \to 0$ as $n \to \infty$, regardless of the path length.

For RLS on monotonic problems, we can manage without the initial λ improvements, as long as we choose $h(t) = ct$ for some suitable constant $c > 1$.

Theorem 3. *The entire runtime termination criterion with* $h(t) = ct$ *succeeds for RLS on monotone problems, with high probability, for* $c > 1$.

Proof. Suppose there have been k improvements so far. We consider three different cases. First, if $1 \leq k \leq n/e$, we have

$$\Pr(\text{fail at stage } k) = \sum_{t=k}^{\infty} \Pr(k \text{ improvements found after time } t)\left(\frac{k}{n}\right)^{ct}$$

$$\leq \sum_{t=k}^{\infty} \Pr(k \text{ improvements found after time } t)\left(\frac{k}{n}\right)^{ck} = \left(\frac{k}{n}\right)^{ck}.$$

Now the function $g(x) = x^x$ has a single minimum in the range $0 < x < 1$ when $x = 1/e$. This means that it decreases on the range $0 < x < 1/e$. Writing $x = k/n$ we see that if $1 \leq k \leq n/e$, we have

$$\left(\frac{k}{n}\right)^{ck} = \left(\left(\frac{k}{n}\right)^{k/n}\right)^{cn} \leq \left(\left(\frac{1}{n}\right)^{1/n}\right)^{cn} = \left(\frac{1}{n}\right)^{c}.$$

In the second case we consider $n/e \leq k \leq n - \ln n$. Now since $g(x) = x^x$ increases on the range $1/e < x < 1$ we have

$$\Pr(\text{fail at stage } k) \leq \left(1 - \frac{\ln n}{n}\right)^{cn - c\ln n} \leq \exp(-c\ln n + c(\ln n)^2/n)$$

$$= \frac{1}{n^c}\exp(c(\ln n)^2/n) \leq \frac{e}{n^c}.$$

The third case is $n - \ln n \leq k \leq n - 1$. Now we have

$$\Pr(\text{fail at stage } k) = \Pr(\text{fail at stage } k \,|\, t < (n\ln n)/4)\,\Pr(t < (n\ln n)/4)$$
$$+ \Pr(\text{fail at stage } k \,|\, t \geq (n\ln n)/4)\,\Pr(t \geq (n\ln n)/4)$$
$$\leq \Pr(t < (n\ln n)/4) + \Pr(\text{fail at stage } k \,|\, t \geq (n\ln n)/4).$$

Since t represents the time taken to attain k improvements, we have (for $n - \ln n \leq k \leq n - 1$):

$$E[t] = n(H_n - H_{n-k}) \geq n(\ln n - \ln\ln n - 1/2) \geq (n\ln n)/2.$$

Therefore

$$\Pr(t < (n\ln n)/4) \leq \Pr(t < E[t] - (n\ln n)/4) \leq \exp\left(-\frac{3}{8\pi^2}(\ln n)^2\right) \leq 1/n^c,$$

for sufficiently large n and any constant c (where we have used Witt's tail bound for sums of geometric variables [5,12]). We also have

$$\Pr(\text{fail at stage } k \,|\, t \geq (n\ln n)/4) \leq (1 - 1/n)^{c(n\ln n)/4} \leq 1/n^{c/4}.$$

Therefore we have, for $n - \ln n \leq k \leq n - 1$,

$$\Pr(\text{fail at stage } k) \leq 1/n^c + 1/n^{c/4}.$$

From the union bound, we have the probability of at least one fail over the entire run is at most

$$\frac{n/e}{n^c} + \frac{e(n - \ln n - n/e)}{n^c} + \frac{\ln n}{n^c} + \frac{\ln n}{n^{c/4}}$$

which tends to zero as $n \to \infty$ as long as $c > 1$.   □

8   Conclusions

We have analysed various schemes for deciding when to terminate an evolutionary algorithm. We have shown that methods based purely on the previous improvement time cannot work satisfactorily. We then introduced a rather artificial criterion, the λ-parallel scheme, which for certain settings of λ, guarantees successful termination with high probability on *any* randomised optimisation heuristic algorithm for any problem provided a given condition on the relative improvement probabilities is satisfied—namely that the improvement probabilities do not increase with time, and that the improvement probability at one stage is not massively larger than that at the next stage. We then considered a more natural approach, based on looking at the runtime so far. This was shown to work successfully for RLS on monotonic and path problems. Whether it works in the more general setting remains an open question.

There is also scope for further work by considering different kinds of algorithms (e.g. population-based algorithms, estimation of distribution algorithms), and different kinds of problems (e.g. multi-objective problems, continuous optimisation problems). Addressing this topic has the potential to not only aid algorithm designers and practitioners, but also ensure computing resource (with its associated environmental impact) is deployed efficiently and effectively.

References

1. Bossek, J., Sudholt, D.: Do additional target points speed up evolutionary algorithms? Theor. Comput. Sci. **950**, 20–38 (2023)
2. Cathabard, S., Lehre, P.K., Yao, X.: Non-uniform mutation rates for problems with unknown solution lengths. In: FOGA 2011: Proceedings of the 11th Workshop Proceedings on Foundations of Genetic Algorithms. ACM (2011)
3. Doerr, B., Doerr, C., Kötzing, T.: Solving problems with unknown solution length at almost no extra cost. Algorithmica **81**, 703–748 (2019)
4. Doerr, B., Rajabi, A., Witt, C.: Simulated annealing is a polynomial-time approximation scheme for the minimum spanning tree problem. Algorithmica **86**, 64–89 (2023)
5. Doerr, B.: Probabilistic tools for the analysis of randomized optimization heuristics. In: Doerr, B., Neumann, F. (eds.) Theory of Evolutionary Computation. NCS, pp. 1–87. Springer, Cham (2020). https://doi.org/10.1007/978-3-030-29414-4_1
6. Doerr, B., Goldberg, L.A.: Drift analysis with tail bounds. In: Schaefer, R., Cotta, C., Kołodziej, J., Rudolph, G. (eds.) PPSN XI. LNCS, vol. 6238, pp. 174–183. Springer, Heidelberg (2021). https://doi.org/10.1007/978-3-642-15844-5_18
7. Einarsson, H., et al.: The linear hidden subset problem for the $(1 + 1)$ EA with scheduled and adaptive mutation rates. Theor. Comput. Sci. **785**, 150–170 (2019)
8. Ghoreishi, S., Clausen, A., Joergensen, B.: Termination criteria in evolutionary algorithms: a survey. In: 9th International Joint Conference on Computational Intelligence, pp. 373–384 (2017)
9. Liu, Y., Zhou, A., Zhang, H.: Termination detection strategies in evolutionary algorithms: a survey. In: GECCO 2018: Proceedings of the Genetic and Evolutionary Computation Conference, pp. 1063–1070. ACM (2018)
10. Lobo, F.G., Bazargani, M., Burke, E.K.: A cutoff time strategy based on the coupon collector's problem. Eur. J. Oper. Res. **286**, 101–114 (2020)
11. Rowe, J.E., Sudholt, D.: The choice of the offspring population size in the $(1, \lambda)$ evolutionary algorithm. Theor. Comput. Sci. **545**, 20–38 (2014)
12. Witt, C.: Fitness levels with tail bounds for the analysis of randomised search heuristics. Inf. Process. Lett. **114**, 38–41 (2014)

Experimental and Theoretical Analysis of Local Search Optimising OBDD Variable Orderings

Thomas Jansen$^{(\boxtimes)}$ and Christine Zarges

Department of Computer Science, Aberystwyth University,
Aberystwyth SY23 3DB, UK
{t.jansen,c.zarges}@aber.ac.uk

Abstract. Building on recent interest in the analysis of the performance of randomised search heuristics for permutation problems we investigate the performance of local search when applied to the classical combinatorial optimisation problem of finding an optimal variable ordering for ordered binary decision diagrams, a data structure for Boolean functions. This brings theory-oriented analysis towards a practically relevant combinatorial optimisation problem. We investigate a class of benchmark functions as well as the leading bit of binary addition, both Boolean functions where the variable ordering makes the difference between linear and exponential size (measured in the number of variables). We present experiments with two local search variants using five different operators for permutations from the literature. These experiments as well as theoretical results show which operators and local search variants perform best, improving our understanding of the operators and local search in combinatorial optimisation.

Keywords: local search · OBDD variable ordering · run time analysis

1 Introduction

Randomized search heuristics are a rich and diverse class of algorithms that are often used for optimization of problems that are hard to solve, where no good problem-specific algorithms are available, and where it is good enough to hope to find a good solution in acceptable time without a formal guarantee concerning run time or solution quality. Evolutionary algorithms [14], ant colony optimisation [11], local search [22], simulated annealing [20] and many others belong to this class of algorithms. They have in common that they perform heuristic search in a space of potential solutions (the *search space*), either maintaining an actual sample of such potential solutions (most heuristics do this) or a probability distribution over the search space (e.g., ant colony optimisation and estimation of

We acknowledge the support of the Supercomputing Wales project, which is part-funded by the European Regional Development Fund via Welsh Government.

© The Author(s), under exclusive license to Springer Nature Switzerland AG 2024
T. Stützle and M. Wagner (Eds.): EvoCOP 2024, LNCS 14632, pp. 177–192, 2024.
https://doi.org/10.1007/978-3-031-57712-3_12

distribution algorithms). The choice and design of the heuristic depends crucially on the search space.

One kind of search space that has been studied to some extent (but arguably less than the space of fixed-length bit strings or fixed-length real-valued vectors) is the space of permutations of n elements, S_n. The probably best-known example for a combinatorial optimisation problem that has S_n as its natural search space is the travelling salesperson problem (TSP) where an order of n cities needs to be found that minimises the travel distance when visiting the cities in this order and returning to the start. There are many other problems where the natural search space is S_n, and we consider the variable ordering problem for ordered binary decision diagrams (OBDDs) here. OBDDs are a data structure for Boolean functions (see Sect. 2 for a description). They provide a compact representation for many important Boolean functions but the size of the data structure depends on the order of variables that appear in the Boolean function that is represented. Finding a good variable ordering is a difficult problem [4] and many different heuristics have been suggested to tackle this problem. We consider local search with five different notions of neighbourhood (implicitly defined by the move operator we apply) that have been suggested before. Different from previous work we are working towards a theoretical analysis, aiming to improve our understanding of heuristic optimisation of permutation problems. We consider a relevant combinatorial optimisation problem and an algorithmic framework (local search with existing operators/neighbourhoods) to make sure that the insights gained are relevant and leading towards a theoretical underpinning of actual practical work. We see our work in the tradition of other work that aimed at bridging the gap between theory and practice in heuristic optimisation [18]. We consider a class of Boolean example functions and a practically relevant Boolean function. For all example functions that we consider a bad ordering implies exponential OBDD size (measured in the number of variables) while a good variable ordering leads to compact OBDDs that are easy to handle even for large numbers of variables. We prove that for some move operators and some functions the induced landscape defines local optima so that local search might fail while for other operators on the same functions no local optima exist and an optimal variable ordering can even be found efficiently, in expected polynomial time. We extend our findings empirically by presenting results of experiments that paint a clear picture of what operators can achieve for those example problems.

We provide important background information and an overview of related work in the following section. Section 3 contains definitions of the Boolean functions and specific local search algorithms we consider. It also contains helpful observations about the functions we consider that support theoretical analysis. Our results are presented in Sect. 4, describing experimental as well as theoretical results. We summarise our results and point out directions for future research in Sect. 5.

2   Background and Related Work

We consider the problem of computing an optimal variable ordering for an OBDD, i.e., finding a variable ordering that leads to the smallest OBDD possible for a given function. Since the satisfiability problem can be solved in time linear in the size of the OBDD if the Boolean function is given as an OBDD, it is clear that the problem of transforming any representation of a Boolean function where the satisfiability problem is hard (like digital circuits or conjunctions of clauses) is hard, too. But even if a Boolean function is given as an OBDD finding a variable ordering that minimises the OBDD size is NP-hard [4]. Since finding an optimal (or at least good) variable ordering is a practically important problem when dealing with OBDDs it comes as no surprise that a number of different heuristics have been considered. Rudell [25] presents a simple greedy heuristic called sifting that is implemented in many OBDD software packages. Bollig, Loebbing and Wegener [3] remark that many heuristics are based on local move operators (essentially leading to local search). They explicitly state that 'the quality of these algorithms cannot be evaluated theoretically' and continue by presenting experimental results of a simulated annealing algorithm based on such local moves. They find better variable orderings than sifting but their algorithm has a larger run time. A significant number of different heuristics for the same problem have been presented, notably a number of different evolutionary algorithms [5,6,12,21]. This includes an evolutionary algorithm implementing a kind of meta approach where the evolutionary algorithm aims to learn good heuristics for the variable ordering problem [13]. Another noteworthy approach using evolutionary algorithms explicitly attempts to find approximate solutions to the problem in order to increase efficiency, employing multi-objective evolutionary optimisation [27]. Other heuristics that have been applied to the problem include scatter search [16] and particle swarm optimisation [1]. All of these approaches that we mention, from the first heuristic 30 years ago [25] to the most recent publication from 2023 [1], have in common that they are purely experimental. But different from 1996 when Bollig, Loebbing and Wegener [3] remarked that 'the quality of these algorithms cannot be evaluated theoretically' there have been very significant advances in our ability to analyse the performance of randomised search heuristics theoretically. While the vast majority of this work has considered the search space of fixed-length Boolean strings [10] there are many results for permutations and a recent increased interest in the development of theory for the performance of randomised search heuristics for permutation problems [9].

The authors of the seminal paper in this area consider the problem of sorting and analyse the performance of simple evolutionary algorithms, concentrating on the question how modelling the problem influences the performance, considering different measures of sortedness of partially sorted sequences as ways to measure the fitness [26]. The same problem has been studied much more recently in the context of noisy optimisation [15]. Other early works include run time analysis of simple evolutionary algorithms for the Eulerian cycle problem [24] but unsurprisingly a lot of effort has been invested in the theoretical study of the run time of evolutionary algorithms on the probably best-known permutation problem, the travelling salesperson problem (TSP). This includes parameterised analysis [28,29], fixed budget analysis [23], analysis of the run time of modern, theory-driven evolutionary algorithms [2] and analysis of evolutionary diversity optimisation [8].

A recent research direction is the development of benchmark functions that are permutation problems to support and facilitate the theoretical study of evolutionary algorithms and other randomised search heuristics for permutation problems [2,9]. The hope is that good benchmark functions help the development of new insights and analytical tools in the area of permutation problems like this has happened for the search space of fixed-length bit strings. Since the study of different combinatorial optimisation problems has also played an important role we propose to study the OBDD variable ordering problem as a different, practically important combinatorial optimisation problem. Different from the TSP it is not invariant to rotations: while for the TSP it makes no difference which city is chosen as first for OBDDs it can make a huge difference if a variable is at the top or bottom of the variable ordering, even if the two different orderings differ only by rotation.

3   Algorithms and Functions

A Boolean function $g \colon \{0,1\}^n \to \{0,1\}$ is defined over Boolean variables x_1, x_2, ..., $x_n \in \{0,1\}$. We call $x \in \{0,1\}^n$ an assignment of values to the variables and $g(x)$ denotes the function value of g for this assignment. We consider subfunctions $g_{|x_i=c}$ (for $i \in \{1, 2, \ldots, n\}$ and $c \in \{0,1\}$) where the value of x_i is fixed to $x_i = c$. We extend this concept to partial assignments $\tilde{x} \in \{0,1,*\}$ where $g_{|\tilde{x}}$ denotes the function where the value of x_i is fixed to the value given by $\tilde{x}$ where this value is different from $*$. We say that g essentially depends on x_i if $g_{|x_i=0} \neq g_{|x_i=1}$.

OBDDs are a fundamental data structure for Boolean functions introduced by Bryant [7]. A πOBDD is a directed graph that represents a Boolean function $g \colon \{0,1\}^n \to \{0,1\}$ that is defined over Boolean variables $x_1, x_2, \ldots, x_n \in \{0,1\}$ with a ordering of these variables π. The graph has exactly one source, i.e., a variable without incoming edges. Its sinks (nodes without outgoing edges) are labelled with either 0 or 1. All other nodes (including the source) are labelled with a variable from $\{x_1, x_2, \ldots, x_n\}$ and have exactly two outgoing edges, one labelled 0 and the other labelled 1. (In diagrams we omit the edge labels and draw the 0-edge as dashed arrow and the 1-edge as solid arrow, see Fig. 1 for an example.) It respects the variable ordering $\pi \in S_n$: on every path from the source to a sink, if a node labelled x_i appears before a node labelled x_j then x_i comes before x_j in π (for any x_i and any x_j). This implies that each variable can appear at most once on any path and that the graph does not contain cycles. For each $x \in \{0,1\}^n$ the function value $g(x)$ of the function g the πOBDD represents can be computed by starting at the source and following a path to a sink by following the 0-edge of the current node that is labelled with x_i if $x_i = 0$ and following the 1-edge otherwise.

The size of a πOBDD is the number of nodes (including the sinks). For each function g and each variable ordering π there is a unique πOBDD of minimal size that can be obtained efficiently (in time linear in the size of the OBDD) by reducing an arbitrary πOBDD by the application of the following two reduction rules in arbitrary order until no further application of those rules is possible. (1) If a variable node has the same 0- and 1-successors then the node can be removed. (2) If two nodes have the same label and the same 0- and 1-successors then the two nodes can be merged (including sinks). The resulting πOBDD is called reduced. In the following we will always work with reduced πOBDDs. In cases where it is clear from the context which variable ordering π is used we may just use OBDD.

We consider the problem of finding a variable ordering π that minimises the size of the reduced πOBDD for all possible variable orderings. Obviously, the space of permutations S_n is the natural search space for this problem. It is known that this problem is NP-hard even if g is given in form of an OBDD [4].

We consider two different kinds of example functions g in this paper. For both the difference between the smallest and largest OBDD is exponential in n, depending on the choice of variable ordering. The first class of example functions is purposefully designed to have this property while being generally simple. We call it $\textsc{DisMon}_k$ (short for 'disjoint monomials of length k'). It is similar in spirit to the so-called real royal road functions for crossover [17].

Definition 1. *For $n, k \in \mathbb{N}$ with $n \geq k$ we define $\textsc{DisMon}_k \colon \{0,1\}^n \to \{0,1\}$ as logical OR of $\lfloor n/k \rfloor$ tuples of variables for which we compute a logical AND:*

$$\textsc{DisMon}_k(x) = \bigvee_{i=0}^{\lfloor n/k \rfloor - 1} \bigwedge_{j=1}^{k} x_{ik+j}.$$

Lemma 1. *Consider $\textsc{DisMon}_k$ for $n, k \in \mathbb{N}$ with $n \geq k$. For each $i \in \{0, 1, \ldots, \lfloor n/k \rfloor - 1\}$, we call the variables $x_{ik+1}, x_{ik+2}, \ldots, x_{ik+k}$ the i-th tuple. For a variable ordering π, we say that the i-th tuple is not interrupted if its variables appear in π without any other variable between them. Otherwise we call the tuple interrupted.*

Any variable ordering without any interrupted tuples is optimal and the corresponding OBDD size is $k \cdot \lfloor n/k \rfloor + 2$. Any variable ordering with interrupted tuples is not optimal and implies larger OBDD size.

Proof. The function depends essentially on the first $k \cdot \lfloor n/k \rfloor$ variables: to see this for variable x_j (with $j \leq k \cdot \lfloor n/k \rfloor$) let i be given such that the i-th tuple contains x_j. Define the partial assignment of values $\{0,1\}$ to variables $\tilde{x}$ by setting all variables in the i-th tuple except x_j to 1 and all other variables in all other tuples to 0. We have $\textsc{DisMon}_k(x)|_{\tilde{x}} = x_j$. Thus, each OBDD for $\textsc{DisMon}_k$ contains at least $k \cdot \lfloor n/k \rfloor + 2$ nodes (one for each essential variable and the two sinks).

For any variable ordering π without interrupted tuples we construct the obvious partial OBDD for the logical AND of the k variables in each tuple. The first node in the first tuple in π is the root of the OBDD. The order of the variables within the tuple is not essential. The 0-successor of each node is the node in the next tuple that comes first in that tuple in π. The 1-successor of the last node in the tuple is the 1-sink. In the final tuple the 0-successor of each node is the 0-sink. This πOBDD has size $k \cdot \lfloor n/k \rfloor + 2$ (one node for each variable in each tuple plus two sinks) and computes DisMon_k.

Now consider an order π' with at least one interrupted tuple. Let this be the i-th tuple. Let x_a and x_b be two variables from the i-th tuple such that a comes before b in π' and there is at least one variable x_c from another tuple between x_a and x_b. Any π'OBDD for DisMon_k contains at least two nodes for x_c: otherwise, we can construct an assignment of values to variables that corresponds to a path in the OBDD that includes x_a that can be extended to include x_c and irrespective of the value of x_a the path includes the same node with label x_c. Now, if we set all variables outside the i-th tuple to 0 and all variables in the i-th tuple to 1 except for x_a then the resulting function equals x_a. However, since we reach the same node with label x_c irrespective of the value of x_a this π'OBDD cannot compute DisMon_k.   □

We remark without proof that any order where some variable has distance $\Omega(n)$ from some other variable of the same tuple implies OBDD size $2^{\Omega(n/k)}$.

The second example function we consider is the leading bit of the result of adding two binary numbers in standard binary encoding. Since addition is a fundamental arithmetic operation representing the leading bit is in this sense a fundamental, practically relevant example function. We consider two numbers between 0 and $2^{(n/2)-1} - 1$ encoded in standard binary encoding using $n/2$ bits each (for even n). The first number is represented using bits $(x_1, x_3, x_5, \ldots, x_{n-1})_2$ and the other $(x_2, x_4, x_6, \ldots, x_n)_2$. The sum is $(c, s_1, s_2, s_3, \ldots, s_{n/2})_2$ and we only consider the most significant bit c. We call the resulting function Add.

Definition 2. *For even $n \in \mathbb{N}$ we define*

$$\text{Add}(x) = \bigvee_{i=1}^{n/2} \left(x_{2i-1} \wedge x_{2i} \wedge \bigwedge_{j=1}^{i-1} (x_{2j-1} \oplus x_{2j}) \right)$$

where we use $\oplus$ for the exclusive or operator.

It is easy to see that the OBDD size for the variable ordering $x_1, x_2, x_3, \ldots, x_{n-1}, x_n$ is $3(n/2)+1$ while it is $2^{\Omega(n)}$ for $x_1, x_3, \ldots, x_{n-1}, x_2, x_4, \ldots, x_n$ (and many other orderings).

Local search is a fundamental randomised search heuristic that aims at optimising an objective function $f: S \to \mathbb{R}$. We will use the OBDD size as our objective function. Local search starts with some initial search point $s \in S$ and operates in the following simple rounds. In each round, it moves to a neighbour of s if that point has a better (or, depending on the specific local search variant, no worse) function value. We can define the neighbourhood $N: S \to 2^S$ that defines to which points $N(s)$ local search can move from the current search point s. We consider the following two popular variants of local search.

Definition 3. *First Improvement Random Local Search (First Improvement RLS) begins with a search point $s \in S$ selected uniformly at random. In each round it selects an order on the neighbourhood $N(s)$ of the current search point s uniformly at random. It considers the neighbours $s' \in N(s)$ in this order and replaces s with the first s' where $f(s') > f(s)$. If no such neighbour s' exists the algorithm terminates.*

It is easy to see that First Improvement RLS terminates after a finite number of steps if S is finite. A point $s \in S$ where First Improvement RLS terminates is called a local optimum. A local optimum s with $f(s) = \max\{f(s') \mid s' \in S\}$ is called global optimum.

An even simpler local search variant is random local search.

Definition 4. *Random Local Search (RLS) begins with a search point $s \in S$ selected uniformly at random. In each round it selects $s' \in N(s)$ uniformly at random. If $f(s') \geq f(s)$ then s' replaces s.*

We define RLS without termination criterion (formally as an infinite random process). While RLS is simpler it is wasteful: it can (and often does) consider the same neighbouring search point multiple times. It is popular because it is similar in structure to a very simple evolutionary algorithm, the (1+1) EA, and often simpler to analyse, making it an attractive starting point for a theoretical analysis. Its wastefulness often makes it moderately slower than First Improvement RLS in locating a local optimum.

The five operators we consider to define five different neighbourhoods have all been considered before [3] and used by a number of different authors in different contexts. They naturally fall into two groups and we describe and discuss them according to this grouping.

Definition 5. *Application of the operator* EXCHANGE(i, j) *(for $i < j \in \{1, 2, \ldots, n\}$) to the variable ordering $\pi = (x_1, x_2, \ldots, x_i, \ldots, x_j, \ldots, x_n)$ results in the variable ordering $(x_1, x_2, \ldots, x_j, \ldots, x_i, \ldots, x_n)$ (exchanging the places of x_i and x_j).*

The operator SWAP(i, j) *is the same as* EXCHANGE(i, j) *but it is only defined for $i \in \{1, 2, \ldots, n - 1\}$ and $j = i + 1$ (exchanging the places of x_i and its successor x_{i+1}).*

The swap operator is important because it is the basis of the popular sifting heuristic for improving the variable ordering of OBDDs [25]. The sifting heuristics starts with the current variable ordering, considers all $n - 1$ different swap operations, picks the best one and repeats this until no further improvement is found. It essentially implements a greedy local search with the swap operator. The exchange operator is the obvious generalisation of the swap operator.

Both, swap and exchange work by exchanging two variables. An alternative is to pick only one variable and move it to a different place. This is done in the move operator and its variants.

Definition 6. *Application of the operator* MOVE(i, j) *(for* $i, j \in \{1, 2, \ldots, n\}$ *to the variable ordering* $\pi = (x_1, x_2, \ldots, x_{i-1}, x_i, x_{i+1}, \ldots, x_n)$ *results in the variable ordering* $(x_1, x_2, \ldots, x_{j-1}, x_i, x_j \ldots, x_{i-1}, x_{i+1}, \ldots, x_n)$ *if* $j < i$ *and* $(x_1, x_2, \ldots, x_{i-1}, x_{i+1}, \ldots, x_{j-1}, x_i, x_j \ldots, x_n)$ *if* $j > i$ *(in both cases moving* x_i *to position* j*).*

The operator MOVE-DOWN(i, j) *is the same as* MOVE(i, j) *but it is only defined for* $j > i$.

The operator MOVE-UP(i, j) *is the same as* MOVE(i, j) *but it is only defined for* $j < i$.

Swap is a restriction of exchange (and also of move) and move-up/move-down are restrictions of move. For local search a restriction can have more but not fewer local optima: local optima for an operator are also local optima for any restriction but not necessarily the other way around. The reason is that a restriction defines a neighbourhood that is a proper subset of the neighbourhood of the more general move operator.

4   Experimental and Theoretical Analysis

We perform experiments for DISMON$_k$ (with $k \in \{2, 3, 4\}$) and ADD(x) for $n \in \{6, 12, 18, \ldots, 60\}$. For each setting, we perform 30 independent runs. As usual we measure run time in terms of function evaluations. For First Improvement RLS we let the algorithm run until it terminates. For RLS we need to decide when to terminate its run. We do so by letting RLS perform n times more function evaluations than First Improvement RLS performed on average for the same function with the same neighbourhood, a generous amount of time. To make results comparable we report the number of function evaluations until the last improvement is found. We present selected results using box plots.

The optimal fitness values, i.e., OBDD sizes, are shown in Fig. 1 (right-hand side). The same figure shows example OBDDs for $n = 6$ (left). For DISMON$_k$ and an optimal variable ordering the OBDD contains the two sinks and one node for each variable, implying size $n + 2$. If k is not a divisor of n we have to replace n with $k \lfloor n/k \rfloor$ and the variables x_i with $i > k \lfloor n/k \rfloor$ are not in the OBDD. For ADD and the optimal variable ordering (which contains the bits of the two numbers that are added in interleaved order of decreasing significance) the OBDD contains for each pair of variables (for each position) three nodes, a complete binary tree over both variables. If both variables have the value 0 the function value is 0. If both variables have the value 1 the function value is 1. If exactly one of the two variables is 1 we continue with the next pair. We can save one node for the final pair (the two least significant bits) because we only need to distinguish if both variables are 1 or not. This implies optimal size $3 \cdot (n/2) - 1 + 2 = (3/2)n + 1$.

Our implementation is in C++ (GNU 10.2.0) and uses the TeDDy library[1] for the creation of OBDDs for a given variable ordering. All experiments are

---

[1] https://michalmrena.github.io/teddy.html.

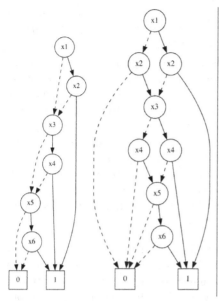

	DisMon$_k$ $(k \cdot \lfloor n/k \rfloor + 2)$			Add(x)
n	$k = 2$	$k = 3$	$k = 4$	$((3/2)n + 1)$
6	8	8	6	10
12	14	14	14	19
18	20	20	18	28
24	26	26	26	37
30	32	32	30	46
36	38	38	38	55
42	44	44	42	64
48	50	50	50	73
54	56	56	54	82
60	62	62	62	91

Fig. 1. Left: Optimal solutions for $n = 6$ for DisMon$_k$ with $k = 2$ (left OBDD) and Add(x) (right OBDD). For Add(x) we compute the most significant bit in the result of $(x_1x_3x_5)_2 + (x_2x_4x_6)_2$. **Right:** Table with optimal fitness values (OBDD sizes).

performed on a high-performance computing cluster with 126 nodes, 384 GB RAM per node and Intel Xeon Gold 6148 CPUs @ 2.40 GHz.

4.1   Results for Swap and Exchange

For swap we begin with the observation that the neighbourhood definition implies the existence of local optima which are not global optima. Therefore, local search can get stuck in such a local optimum and fail to find an optimal solution.

Theorem 1. *First Improvement RLS with exchange or swap neighbourhood does not have a finite expected optimisation time for* DisMon$_k$. *It fails to find a global optimum with positive probability.*

Proof. We prove that there is a local optimum for exchange. Remembering that swap is a restricted version of exchange (any swap operation is also a valid exchange operation but there are exchange operations that are not swaps) we see that any local optimum for exchange is also a local optimum for swap. Thus, it suffices to prove the statement for exchange. We assume $n/k \in \mathbb{N}$.

The first statement is a direct consequence of the second. Remember that we use a random permutation as starting point. Therefore, any permutation appears as first search point with positive probability (bounded below by $1/(n!)$).

Consider the permutation $x_2, x_3, x_4, \ldots, x_{n-1}, x_n, x_1$. For any k, only the first k-tuple is interrupted in this variable ordering. The first k-tuple is split with x_1 missing. This implies that the size of the OBDD is larger than the optimal size. Consider any exchange operation that leads from a non-optimal variable ordering to an optimal variable ordering. This operation needs to decrease the number of interrupted tuples from a positive value to 0. It cannot decrease the number from 1 to 0: if the two variables that are exchanged belong to the same tuple then the number of interrupted tuples does not change. If the two variables that are exchanged belong to different tuples then either the number of interrupted tuples does not change or it increases. Since we start with exactly one interrupted tuple there is no exchange operation leading to an optimal variable ordering without first increasing the number of interrupted tuples but such a move will not be accepted by First Improvement RLS. Thus, the variable ordering above is a local optimum for exchange.   □

We perform experiments for the swap and exchange operators. We observe that swap performs badly on all considered functions and First Improvement RLS. While Theorem 1 indicates that this might happen we now see confirmation that it occurs with high probability. We depict DISMON_k with $k = 2$ as one example in Fig. 2(a). Even the best variable orderings swap finds can be very bad. This demonstrates the importance of finding a good variable ordering.

The exchange operator performs much better in experiments and finds an optimal variable ordering for DISMON_k with $k = 2$ and $k = 3$ in all 30 runs. However, for $k = 4$ and $\text{ADD}(x)$ the performance deteriorates considerably with increasing n. While for DISMON_k with $k = 4$ still at least one of the 30 runs reaches an optimal solution (see Fig. 2(b)), exchange is unable to find an optimal solution for $\text{ADD}(x)$ if $n > 30$. This demonstrates that the local optimum that exists for the swap and exchange operators for DISMON_2 is likely to be encountered when using the swap operator but much more likely to be avoided when using the exchange operator.

We also consider experiments for RLS and see that the performance of swap is improved considerably in comparison to First Improvement RLS. DISMON_k with $k = 2$ and $k = 3$ are optimised in all 30 runs. For $k = 4$ and $\text{ADD}(x)$ the local optima found have much better fitness if RLS is used. For example, for $n = 30$, we observe values between 30 and 146 for RLS on DISMON_k with $k = 4$ while First Improvement RLS achieved values between 141 and 792. Similar results can be observed for $\text{ADD}(x)$. The main difference between RLS and First Improvement RLS with respect to solution quality is that RLS accepts moves to search points with equal fitness, allowing it to perform a random walk of plateaus of equal quality. We see that this is crucial for the swap operator: to achieve a fitness improvement several consecutive swaps can be needed.

For the exchange neighbourhood we also observe improved performance: RLS with this operator is able to optimise DISMON_k with $k = 4$ in all 30 runs. For $\text{ADD}(x)$ we also see performance deteriorating as n grows. However, the fitness achieved by RLS (Fig. 2(d)) is slightly better than the fitness achieved by First Improvement RLS (Fig. 2(c)). For example for $n = 60$, the fitness of First

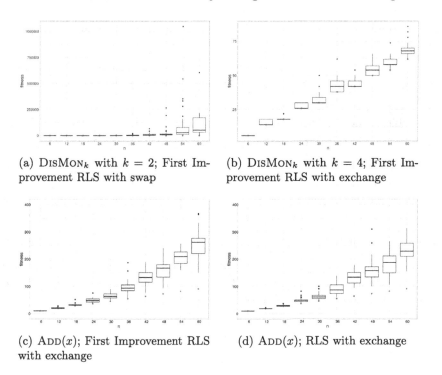

(a) DisMon_k with $k = 2$; First Improvement RLS with swap

(b) DisMon_k with $k = 4$; First Improvement RLS with exchange

(c) $\text{Add}(x)$; First Improvement RLS with exchange

(d) $\text{Add}(x)$; RLS with exchange

Fig. 2. Experimental results for swap and exchange. The optimal value for each value of n is shown as a red asterisk $*$. (Color figure online)

Improvement RLS is between 151 and 367 with an average of 261.1 while for RLS the fitness is between 154 and 313 with an average of 232.7, possibly due to the ability to perform a random walk on a plateau of equal fitness.

4.2   Results for Move

We again perform experiments as described above to investigate how the three variants of move perform on our example instances if First Improvement RLS is used. We observe that move clearly outperforms move-up and move-down on all considered functions with the sole exception of DisMon_2 (i.e., DisMon_k with $k = 2$ where we only consider pairs but not tuples of bigger size). Move finds an optimal solution for DisMon_k with $k = 2$ and $k = 3$ in all 30 runs. It locates an optimal solution in at least one run for $k = 4$ but is unable to optimise $\text{Add}(x)$ for larger values of n. We show examples for DisMon_k with $k = 4$ and $\text{Add}(x)$ in Figs. 3(a) and 3(b), respectively.

The striking success of all three variants of the move operator in DisMon_k for $k = 2$ in comparison with the failure of the swap operator can be explained theoretically. While we had no finite expected optimisation time for First Improvement RLS with swap neighbourhood we can prove that First Improvement RLS has polynomial expected optimisation time for any of the three move neighbourhoods on DisMon_2.

Theorem 2. *First Improvement RLS with any move neighbourhood (move, move-up, move-down) optimises DisMon_2 in expected polynomial time.*

Proof. For a given variable ordering π and variable x_i let $u_\pi(x_i)$ be the number of variables appearing before x_i in π that have the other variable in their pair appear behind x_i in π. The πOBDD size of DisMon_2 is $2 + \sum_{i=1}^{n} 2^{u_\pi(x_i)}$. A variable ordering π is optimal if and only if $u_\pi(x_i) = 0$ for all x_i. For any π and x_i with $u_\pi(x_i) > 0$ there is a move operation that decreases this value by 1 by moving the other variable in the pair before x_i (move-down) or x_i after this variable (move-up). The probability to decrease the largest $u_\pi(x_i)$ value by 1 is $\Omega(1/n^2)$ since there are less than n^2 move operations. Since $2 + \sum_{i=1}^{n} 2^{u_\pi(x_i)}$ is the objective function value and therefore strictly decreasing we see that after $O(n^2)$ such move operations we have $u_\pi(x_i) = 0$ for all x_i. This establishes $O(n^4)$ as upper bound on the expected optimisation time. $\square$

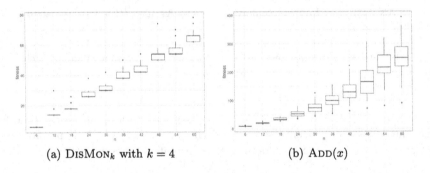

(a) DisMon_k with $k = 4$ (b) $\text{Add}(x)$

Fig. 3. Experimental results for First Improvement RLS with move. The optimal value for each value of n is shown as a red asterisk $*$. (Color figure online)

Looking at move-up and move-down we see that both operators again optimise DisMon_k with $k = 2$, however, the performance for $k = 3$ decreases considerably. We show move-up as an example in Fig. 4 and observe that First Improvement RLS with move-up neighbourhood still manages to find an optimal solution in at least one of the 30 runs but not in all 30 runs as the move operator does. The results for move-down are similar (omitted due to limited space).

Finally, we consider RLS with move neighbourhood and observe that in the same way as with exchange neighbourhood RLS is able to optimise DisMon_k with $k = 4$ in all 30 runs. Similarly, RLS with move neighbourhood exhibits similar behaviour to First Improvement RLS with the same move operator on $\text{Add}(x)$.

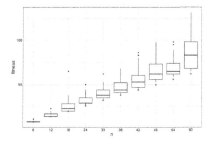

Fig. 4. Experimental results for DisMon$_k$ with $k = 3$ using First Improvement RLS with move-up. The optimal value for each value of n is shown as a red asterisk $*$. (Color figure online)

4.3   Average Run Times

In cases where different heuristics find solutions of different quality a heuristic with poorer performance might still be preferable if it runs much faster. Already Bollig, Loebbing, and Wegener [3] mention their approach finds much better variable orderings than sifting but the much higher speed of sifting might make it the preferred choice for some applications. However, when two different heuristics perform essentially equal in terms of solution quality there is no question that one will prefer the faster heuristic. Therefore, we consider the average run times of First Improvement RLS with exchange, move, move-up and move-down neighbourhood for DisMon$_k$ with $k = 2$: for all four operators the optimum variable ordering is consistently found. As usual in theoretical analysis we use the number of function evaluations to measure time. While this measure is obviously less intuitive in comparison to wall clock time it has the advantage of being much more robust with respect to implementation details and hardware platform. Also, it is known that bringing theoretical analysis closer to actual time is a challenging separate research area [19].

For comparison we present box plots of the average first hitting times of the global optimum: we consider the number of function evaluations until a global optimum is found for the first time and look at averages over 30 runs. We ignore the time it takes First Improvement RLS to terminate (which happens when no better search point can be found) to make the results comparable with RLS which does not terminate. Comparing these average run times for the four operators (Figs. 5(a)–5(d)) we see that the move operator is clearly the slowest and its two restricted variants, move-up and move-down, are fastest. The exchange operator's performance is between the performance of move and its two variants.

If we perform experiments for RLS we see performance very similar to First Improvement RLS with move and exchange neighbourhoods, and this is true for all operators. We omit the actual results due to space restrictions.

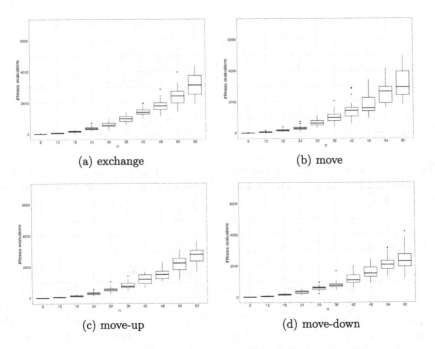

Fig. 5. First hitting times for First Improvement RLS on DisMon₂, different operators.

5   Conclusions and Future Work

We have investigated empirically and theoretically the performance of two different local search variants with five different neighbourhoods for the combinatorial optimisation problem of finding an optimal variable ordering for OBDDs, considering a class of example functions as well as the leading bit of addition of binary numbers. We consider this to be a step towards a better understanding of the performance of randomised search heuristics on permutation problems.

We have shown that of the five considered neighbourhoods swap performs clearly worst and, in general, move performs better than exchange. Looking at the move neighbourhood we see that the more restricted variants move-up and move-down are better with respect to expected run time if they are sufficient. When comparing First Improvement RLS with RLS we see that the ability of RLS to perform a random walk on plateaus of equal solution quality can be important for solution quality but it can increase run times.

There are a lot of open questions, including tight bounds for the expected optimisation time on ADD and DisMon$_k$. Gaining a better understanding of how and when RLS outperforms First Improvement RLS and a stronger theoretical underpinning by proving more analytical results are at the front of the queue. Another important aspect that we have not considered at all is taking into account the time it takes to construct the OBDDs, something that is required to evaluate the quality of a variable ordering. This time depends on the size of the OBDD that needs to be constructed and there is the option of cutting down on this time by stopping the computation early when a comparison of function values is sufficient and the absolute sizes are not needed.

References

1. Awad, A., Hawash, A., Abdalhaq, B.: A genetic algorithm (GA) and swarm based binary decision diagram (BDD) reordering optimizer reinforced with recent operators. IEEE Trans. Evol. Comput. **27**(3), 535–549 (2023). https://doi.org/10.1109/TEVC.2022.3170212
2. Bassin, A.O., Buzdalov, M.: The $(1 + (\lambda, \lambda))$ genetic algorithm for permutations. In: Coello, C.A.C. (ed.) GECCO 2020: Genetic and Evolutionary Computation Conference, Companion Volume, pp. 1669–1677. ACM (2020). https://doi.org/10.1145/3377929.3398148
3. Bollig, B., Löbbing, M., Wegener, I.: On the effect of local changes in the variable ordering of ordered decision diagrams. Inf. Process. Lett. **59**(5), 233–239 (1996). https://doi.org/10.1016/0020-0190(96)00119-6
4. Bollig, B., Wegener, I.: Improving the variable ordering of OBDDs is NP-complete. IEEE Trans. Comput. **45**(9), 993–1002 (1996). https://doi.org/10.1109/12.537122
5. Brudaru, O., Ebendt, R., Furdu, I.M.: Optimizing variable ordering of BDDs with double hybridized embryonic genetic algorithm. In: Ida, T., Negru, V., Jebelean, T., Petcu, D., Watt, S.M., Zaharie, D. (eds.) 12th International Symposium on Symbolic and Numeric Algorithms for Scientific Computing (SYNASC 2010), pp. 167–173. IEEE Computer Society (2010). https://doi.org/10.1109/SYNASC.2010.33
6. Brudaru, O., Rotaru, C., Furdu, I.M.: Static segregative genetic algorithm for optimizing variable ordering of ROBDDs. In: Wang, D., et al. (eds.) 13th International Symposium on Symbolic and Numeric Algorithms for Scientific Computing (SYNASC 2011), pp. 222–229. IEEE Computer Society (2011). https://doi.org/10.1109/SYNASC.2011.54
7. Bryant, R.E.: Graph-based algorithms for Boolean function manipulations. IEEE Trans. Comput. **35**(8), 677–691 (1986)
8. Do, A.V., Guo, M., Neumann, A., Neumann, F.: Analysis of evolutionary diversity optimization for permutation problems. ACM Trans. Evol. Learn. Optim. **2**(3), 11:1–11:27 (2022). https://doi.org/10.1145/3561974
9. Doerr, B., Ghannane, Y., Brahim, M.I.: Towards a stronger theory for permutation-based evolutionary algorithms. In: Fieldsend, J.E., Wagner, M. (eds.) Genetic and Evolutionary Computation Conference (GECCO 2022), pp. 1390–1398. ACM (2022). https://doi.org/10.1145/3512290.3528720
10. Doerr, B., Neumann, F.: Theory of Evolutionary Computation: Recent Developments in Discrete Optimization. Springer, Cham (2020). https://doi.org/10.1007/978-3-030-29414-4
11. Dorigo, M., Stützle, T.: Ant Colony Optimzation. MIT Press, Cambridge (2004)
12. Drechsler, R., Göckel, N., Becker, B.: Genetic algorithm for variable ordering of OBDDs. IEEE Proc. Comput. Digit. Tech. **143**(6), 364–368 (1996). https://doi.org/10.1049/ip-cdt:19960789
13. Drechsler, R., Göckel, N., Becker, B.: Learning heuristics for OBDD minimization by evolutionary algorithms. In: Voigt, H., Ebeling, W., Rechenberg, I., Schwefel, H. (eds.) PPSN 1996. LNCS, vol. 1141, pp. 730–739. Springer, Cham (1996). https://doi.org/10.1007/3-540-61723-X_1036

14. Eiben, A.E., Smith, J.E.: Introduction to Evolutionary Computing. Springer, Heidelberg (2015). https://doi.org/10.1007/978-3-662-44874-8
15. Gavenciak, T., Geissmann, B., Lengler, J.: Sorting by swaps with noisy comparisons. Algorithmica **81**(2), 796–827 (2019). https://doi.org/10.1007/s00453-018-0429-2
16. Hung, W.N.N., Song, X.: BDD variable ordering by scatter search. In: 19th International Conference on Computer Design (ICCD 2001), pp. 368–373. IEEE (2001). https://doi.org/10.1109/ICCD.2001.955053
17. Jansen, T., Wegener, I.: Real royal road functions – where crossover provably is essential. Discrete Appl. Math. **149**, 111–125 (2005)
18. Jansen, T., Wiegand, R.P.: Bridging the gap between theory and practice. In: Yao, X., et al. (eds.) PPSN 2004. LNCS, vol. 3242, pp. 61–71. Springer, Cham (2004). https://doi.org/10.1007/978-3-540-30217-9_7
19. Jansen, T., Zarges, C.: Analysis of evolutionary algorithms: from computational complexity analysis to algorithm engineering. In: Foundations of Genetic Algorithms (FOGA 2011), pp. 1–14. ACM Press (2011)
20. Laarhoven, P.J.M., Aarts, E.H.L.: Simulated Annealing: Theory and Applications. Springer, Dordrecht (1987). https://doi.org/10.1007/978-94-015-7744-1
21. Lenders, W., Baier, C.: Genetic algorithms for the variable ordering problem of binary decision diagrams. In: Wright, A.H., Vose, M.D., Jong, K.A.D., Schmitt, L.M. (eds.) FOGA 2005. LNCS, vol. 3469, pp. 1–20. Springer, Heidelberg (2005). https://doi.org/10.1007/11513575_1
22. Michiels, W., Korst, J., Aarts, E.: Theoretical Aspects of Local Search. Springer, Heidelberg (2007). https://doi.org/10.1007/978-3-540-35854-1
23. Nallaperuma, S., Neumann, F., Sudholt, D.: Expected fitness gains of randomized search heuristics for the traveling salesperson problem. Evol. Comput. **25**(4), 673–705 (2017). https://doi.org/10.1162/evco_a_00199
24. Neumann, F.: Expected runtimes of evolutionary algorithms for the Eulerian cycle problem. Comput. Oper. Res. **35**(9), 2750–2759 (2008). https://doi.org/10.1016/j.cor.2006.12.009
25. Rudell, R.: Dynamic variable ordering for ordered binary decision diagrams. In: Lightner, M.R., Jess, J.A.G. (eds.) Proceedings of the 1993 IEEE/ACM International Conference on Computer-Aided Design, pp. 42–47. IEEE Computer Society/ACM (1993). https://doi.org/10.1109/ICCAD.1993.580029
26. Scharnow, J., Tinnefeld, K., Wegener, I.: The analysis of evolutionary algorithms on sorting and shortest paths problems. J. Math. Model. Algorithms **3**(4), 349–366 (2004). https://doi.org/10.1007/s10852-005-2584-0
27. Shirinzadeh, S., Soeken, M., Große, D., Drechsler, R.: An adaptive prioritized ε-preferred evolutionary algorithm for approximate BDD optimization. In: Bosman, P.A.N. (ed.) Genetic and Evolutionary Computation Conference (GECCO 2017), pp. 1232–1239. ACM (2017). https://doi.org/10.1145/3071178.3071281
28. Sutton, A.M., Neumann, F.: A parameterized runtime analysis of evolutionary algorithms for the Euclidean traveling salesperson problem. In: Hoffmann, J., Selman, B. (eds.) Proceedings of the Twenty-Sixth AAAI Conference on Artificial Intelligence, pp. 1105–1111. AAAI Press (2012). https://doi.org/10.1609/aaai.v26i1.8273
29. Sutton, A.M., Neumann, F., Nallaperuma, S.: Parameterized runtime analyses of evolutionary algorithms for the planar Euclidean traveling salesperson problem. Evol. Comput. **22**(4), 595–628 (2014). https://doi.org/10.1162/EVCO_a_00119

Author Index

© The Editor(s) (if applicable) and The Author(s), under exclusive license
to Springer Nature Switzerland AG 2024
T. Stützle and M. Wagner (Eds.): EvoCOP 2024, LNCS 14632, p. 193, 2024.
https://doi.org/10.1007/978-3-031-57712-3

Printed in the United States
by Baker & Taylor Publisher Services